Hendrik Sturm

AF514243

Stabilisation behaviour of cyclically loaded shallow foundations for offshore wind turbines

Titelbild Mit freundlicher Genehmigung der Firma Vestas Wind Systems A/S.

Stabilisation behaviour of cyclically loaded shallow foundations for offshore wind turbines

by
Hendrik Sturm

universitätsverlag karlsruhe

Dissertation, Universität Karlsruhe (TH)
Fakultät für Bauingenieur-, Geo- und Umweltwissenschaften, 2009
Hauptreferent: o. Prof. em. Dr.-Ing. Dr. h.c. Gerd Gudehus
Korreferent: o. Univ. Prof. Dipl.-Ing. Dr. Dimitrios Kolymbas

Impressum

Universitätsverlag Karlsruhe
c/o Universitätsbibliothek
Straße am Forum 2
D-76131 Karlsruhe
www.uvka.de

SOME RIGHTS RESERVED

Dieses Werk ist unter folgender Creative Commons-Lizenz
lizenziert: http://creativecommons.org/licenses/by-nc-nd/3.0/de/

Universitätsverlag Karlsruhe 2009
Print on Demand

ISBN: 978-3-86644-413-3

Stabilisation behaviour of cyclically loaded shallow foundations for offshore wind turbines

Zur Erlangung des akademischen Grades eines

DOKTOR-INGENIEURS

von der Fakultät für

Bauingenieur-, Geo- und Umweltwissenschaften
der Universität Fridericiana zu Karlsruhe (TH)

genehmigte

DISSERTATION

von

Dipl.-Ing. Hendrik Sturm

aus Pforzheim

Tag der mündlichen Prüfung: 23. Juli 2009

Hauptreferent: o. Prof. em. Dr.-Ing. Dr. h.c. Gerd Gudehus

Korreferent: o. Univ.-Prof. Dipl.-Ing. techn. Dr. habil. Dimitrios Kolymbas

Karlsruhe, 2009

Kurzfassung

Die Bestrebungen vieler Regierungen einen ausgewogenen Energiemix aus unterschiedlichen Primärenergien zu erzielen, hat in der jüngeren Vergangenheit zu einer Intensivierung der Förderung von erneuerbaren Energiequellen geführt. Verstärkt wird diese durch die aktuellen Diskussionen über Klimaschutz und globaler Erwärmung. Besonders profitieren davon offshore, das heißt im Meer, gegründete Windkraftanlagen. Speziell für diese wurde eine Vielzahl unterschiedlicher, öffentlich und privatwirtschaftlich finanzierter, Forschungs- und Entwicklungsprogramme aufgelegt.

Die derzeit geplanten Anlagen sehen Turbinen mit einer Leistung von mindestens 5 MW vor. Dies bedingt hohe und wind-exponierte Bauwerke mit beachtlichen Rotor-Spannweiten. Die daraus resultierenden großen zyklisch horizontalen Belastungen aus Wind – und für größere Wassertiefen zusätzlich aus Wellen – stellen dabei eine besondere Herausforderung bei der Dimensionierung der Gründung dar.

Eine mögliche Gründungsvariante ist das Flachfundament. Dieses muss die zyklisch horizontalen Lasten und die daraus resultierenden großen Momente an Geländeoberkante, infolge des hohen Lastangriffspunktes, *sicher* in den Baugrund ableiten. Kontinuierliche zyklische Wechsellasten mit konstanten Amplituden führen im Allgemeinen zu einer akkumulierten Setzung; Schwelllasten zusätzlich zu einer akkumulierten Verdrehung des Bauwerkes. In-situ Lasten sind jedoch stark veränderlich in Amplitude und Mittelwert über die Zeit. Deshalb ist es wichtig die Variation der Lasten im Design zu berücksichtigen.

Kleinmassstäbliche Modellversuche an einem Flachfundament, abwechselnd belastet mit Zyklenpaketen großer und kleiner Lastamplitude, haben gezeigt, dass die akkumulierte Verdrehung infolge der größeren Last teilweise wieder kompensiert wird in Phasen mit Schwell- oder Wechsellasten kleinerer Amplitude. D.h. dass sich ein verdrehtes Fundament teilweise wieder aufrichten kann. Numerische Berechnungen zeigen, dass Amplitude, Last-Richtung und Anzahl der Wiederholungen der kleineren Last nur quantitativen Einfluß haben. Das qualitative Verhalten der Rückdrehung ist davon unabhängig.

Die Setzungsrate[1] eines zyklisch belasteten Fundaments hängt vom Zustand des Bodens ab. Die Rate nimmt ab mit kleiner werdender mittlerer deviatorischer Spannung und zunehmender

[1] das ist die Setzung pro Zyklus

Dichte und mittlerem isotropen Druck. Der Zustand des Bodens unter einem verdrehten Fundament infolge einer größeren Belastung, z.B. ein starker Sturm oder ein Soliton, ist inhomogen. D.h. dass auf der Entlastungsseite die deviatorische Spannung größer und Dichte und mittlerer Druck kleiner sind als auf der Kompressionsseite. Deshalb kommt es infolge der anschliessenden zyklischen Belastung mit kleinerer Amplitude und Mittelwert zu einer Rückdrehung des Fundamentes.

Die vertikale Ausrichtung eines Bauwerkes ist notwendig für die Einhaltung der Gebrauchstauglichkeit. Daher ist es angemessen die Rückdrehung als *Selbstheilung* zu bezeichnen. Ferner ist es gerechtfertigt den Mechanismus *Stabilisierung* zu nennen, da die Rate der Rückdrehung kontinuierlich mit zunehmender Zyklenzahl abnimmt. Ein stabiler Zustand ist gekennzeichnet durch ein (lokales) Energieminimum wohingegen instabile Zustände durch eine Beschleunigung, d.h. eine zunehmende Verformungsrate, identifiziert werden können.

Das Stabilisierungsverhalten von Flachgründungen ist nicht eine allgemeine Eigenschaft von Fundamente. Voraussetzungen für eine Stabilisierung sind kleine Verdrehungen und ein tiefer Massenschwerpunkt des Bauwerkes, bezogen auf die Fundamentbreite. Auch der Bodenzustand und die Fundamentgeometrie haben einen grossen Einfluß auf das Stabilisierungsverhalten. So zeigt sich, dass aufgelöste Flachfundamente mit zusätzlichen Schürzen vorteilhaft sind.

Die Arbeit beschreibt mit Hilfe von Modellversuchen und numerischen Berechnungen ein neuartiges und bislang unbekanntes Verhalten von Flachgründungen unter typischen Offshore Bedingungen. In einer qualitativen numerischen Parameterstudie wird der Einfluß unterschiedlicher Größen auf diese Verhalten untersucht. Es wird sowohl ein bodenmechanisches als auch physikalisches Erklärungsmodell vorgestellt. Darauf aufbauend wird eine verbesserte Gründungsvariante vorgeschlagen, die das beschriebene Verhalten vorteilhaft ausnutzt.

Abstract

Many governments endeavour to achieve a balanced energy mix from different primary energies, which has caused in the recent years a significantly increased promotion of renewable energy sources. This is additionally encouraged by the current discussions about greenhouse effect and global warming. In particular offshore founded wind turbines benefit from these developments, because many research and development programs have been established which mainly focus on offshore wind energy.

At the present, wind turbine capacities of at least 5 MW are planned to install. They require tall and wind-exposed structures as well as wide wingspans of the rotor blades. This leads to large cyclic horizontal forces from wind and waves, which pose a particular challenge for the design of the substructure.

One possible solution to found an offshore wind turbine is a shallow foundation. It must be able to transfer the cyclic horizontal loads and the resulting large moments at ground surface, due to the high loading point of the horizontal forces, *safe* into the subsoil. Continuous symmetric load cycles with constant amplitude cause accumulated settlement, while asymmetric load cycles cause in addition an accumulated rotation of the structure. However, in-situ loads are not constant but rather highly fluctuating in time, with respect to amplitude and average value. Thus it is necessary to take into account the variation of loading conditions in the design.

Small-scale model tests on a shallow foundation, subjected alternately to cyclic loads with large and small amplitudes, have shown, that the accumulated rotation due to large amplitudes get partially compensated during phases with smaller amplitudes. That means, that a tilted foundation due to an intensive loading, straightens up again during moderate cyclic loading. Numerical simulations have revealed, that this behaviour of cyclically loaded shallow foundations is quantitatively influenced by the amplitude, relative loading direction and number of load cycles.

The rate of settlement[2] of a cyclically loaded shallow foundation depends on the state of the subsoil. The rate decreases for decreasing average deviatoric stress and increasing mean pressure and density. The state of the soil under a tilted foundation, due to an extreme load event, e.g. a severe storm or a soliton, is inhomogeneous. I.e. the deviatoric stress is larger

[2]that is the settlement per cycle

and the density and mean pressure are smaller on the uplift side than on the compression side. Hence, the foundation rotates back during subsequent cyclic loading.

Since the vertical alignment of a structure is necessary for its serviceability, it is appropriate to denote the back-rotation *self-healing*. The term *stabilisation* is justifiable to describe the mechanism, since the rate of back-rotation decreases with increasing number of load reversals. A stable state is characterised by a (local) minimum of the free energy, whereas unstable states can be identified by an acceleration of the system.

The stabilisation behaviour of shallow foundations is not a general property of substructures. It is restricted to small rotations and to structures with a low centre of gravity, referred to the width of the foundation. Also the initial state of the soil previous to the installation and the shape of the foundation have an important influence on the stabilisation behaviour, namely on the rates of settlement and back-rotation.

This work presents by means of model tests and numerical simulations a novel and up to now unknown behaviour of shallow foundations subjected to typical offshore loading conditions. With the aid of a numerical parameter study, the influences of different parameters are analysed. A soil mechanical as well as a physical explanation model is presented. Based on this, an improved foundation geometry is proposed.

Danksagung

Diese Arbeit ist aus einer Forschungstätigkeit am Institut für Bodenmechanik und Felsmechanik (IBF) an der Universität Karlsruhe hervorgegangen. Das erfolgreiche Abschliessen dieser war jedoch nur möglich durch die Unterstützung vieler Personen und Institutionen bei denen ich mich bedanken möchte.

Meinem besonderen Dank gilt den beiden Referenten Herrn Prof. G. Gudehus und Herrn Prof. D. Kolymbas. Nur aufgrund des mir gegenüber erbrachten Vertrauens durch Herrn Prof. G. Gudehus war es überhaupt möglich das ich an diesem Forschungsprojekt arbeiten konnte. Aus einer Vielzahl ergiebiger und inspirierender Gesprächen mit ihm konnte ich viele Anregungen und Gedanken mitnehmen die massgeblich zum gelingen der Arbeit beitrugen. Wertvolle Hilfe leistete mir Herr Prof. D. Kolymbas beim erlernen des wissenschaftlich strukturierten Arbeitens. Seiner unermüdlichen Geduld ist es zu verdanken, dass diese Arbeit in eine lesbare Form gebracht wurde. Seine stets kritischen Fragen waren sehr hilfreich neue Sichtweisen und Perspektiven aufzuzeigen.

Meinen Dank gilt weiter meinen beiden Arbeitskollegen Herrn Dr. habil. A. Niemunis, der mir in vielen Gesprächen wertvolle Hilfe insbesondere zum Verständnis der mathematischen Formulierung von Stoffgesetzten geleistet hat, sowie Herr H. Wienbroer, ohne dessen Hilfe die Versuche nicht hätten durchgeführt werden können.

Der Firma Ed. Züblin AG aus Stuttgart sei gedankt, da sie den initialen Anstoß für diese Arbeit gegeben hat, und dem Institut für Bodenmechnanik und Felsmechanik (IBF) an der Universität Karlsruhe, dass es diesen Aufgriff und zu einem durch das Bundesministerium für Umwelt, Naturschutz und Reaktorsicherheit (BMU) finanziertes Forschungsprojekt geführt hat, aus dessen Mittel der Autor finanziert wurde.

Weitere finanzielle Unterstützung bekam der Autor durch ein Stipendium des Karlsruhe House of Young Scientists (KHYS), das einen drei-monatigen Aufenthalt am Norges Geotekniske Institutt (NGI) in Oslo, Norwegen, ermöglicht hat. Die dort geführten Gespräche, insbesondere mit Herrn Dr. H.-P. Jostad, waren wertvoll ein besseres Verständnis über die behandelte Materie zu erlangen.

Ferner möchte ich Frau Dr. C. Lang vom Institut für Hydromechanik (IfH) an der Universität Karlsruhe danken, dass Sie mir in der finalen Phase der Arbeit wertvolle Unterstützung leistete.

Und natürlich möchte ich mich bei meiner Familie bedanken, die viel Zeit mit mir entbehren musste damit ich diese Arbeit zu einem erfolgreichen Ende führen konnte.

Karlsruhe, Juli 2009 *Hendrik Sturm*

Contents

Kurzfassung iii

Abstract v

Danksagung vii

1 Introduction 1

1.1 Background 1

1.2 Scope of this work 3

2 Foundation types and design for Offshore Wind Turbines 5

2.1 Locations and site conditions 5

2.2 Foundation types for offshore wind turbines 5

2.2.1 Gravity base structures 5

2.2.2 Bucket foundation 8

2.2.3 Monopile foundation 9

2.2.4 Tripod foundation 10

2.2.5 Jacket foundation 10

2.3 Design methods 11

2.3.1 Installation 12

2.3.2 In-service performance 13

2.3.3 Scour protection 17

2.4 Load assumptions and design loads 17

3 Constitutive models 19

3.1 Hypoplastic models and extensions 19

3.1.1 Hypoplasticity 19

3.1.2 Visco-hypoplasticity 23

3.1.3 Intergranular strain 31

3.2 SaniSand 34

3.3 High cycle accumulation model 39

4 Model tests **45**

4.1 Model foundation 45

4.2 Test rig and instrumentation 46

4.2.1 Instrumentation 47

4.2.2 Model sand and preparation method 48

4.3 Loading program and evaluation procedure 48

4.4 Results of the model tests 50

4.4.1 Measured rotations 50

4.4.2 Measured settlement 52

5 FE-simulations of a substructure founded on one large plate **55**

5.1 Model description 55

5.1.1 FE-Models 55

5.1.2 Constitutive parameters 58

5.2 Back-calculation of model tests 59

5.3 Parameter study 62

5.3.1 Rotational behaviour 64

5.3.2 Settlement behaviour 68

5.3.3 Summary 70

5.3.4 Influence of drainage 71

5.4 Comparison of different constitutive models 73

6 Discussion of the stabilisation behaviour **75**

6.1 (Differential) Settlement 75

6.2 Two explanatory models 81

6.2.1 Soil mechanical approach 81

6.2.2 Physical approach 83

6.3 Pore water and drainage 85

6.4 A visco-hypoplastic description of the self-healing 86

6.5 Concluding remarks 88

7 FE-simulations of a substructure founded on multiple plates **89**

7.1 Structural concept 91

7.1.1 Platforms without skirts 92

7.1.2 Platforms with skirts 93

7.2 FE-model 94
7.2.1 Prediction of model tests 97
7.2.2 Parameter study 100
7.3 Conclusion and Outlook 104

8 Summary and outlook **105**
8.1 Summary 105
8.2 Outlook 106

A Design of the model foundation **107**
A.1 Background 107
A.2 Determination of the diameter of the foundation 108
A.2.1 Bearing capacity 108
A.2.2 Limiting of the eccentricity 110
A.2.3 Prototype and model dimensions 111

B Detailed views of the model test rig **113**

C Load table of a representative Offshore Wind Park **115**

D Convergence study **117**

E Bibliography **119**

1. Introduction

Engineers are regularly faced with the problem to predict the behaviour of foundations subjected to cyclic loading. The causes of alternating and dynamic loads are manifold. They can originate from natural forces, such as wind, sea waves, currents, earthquakes, etc., or operational forces such as traffic, machinery, alternate filling and discharging of tanks and basins, blasting, pile driving, etc..

Several different calculation procedures have been developed to account for the elusive and complex cyclic loads. They depend on the quantity, i.e. on the frequency of appearance and intensity, as well as on the special conditions and requirements of subsoil and building. Cyclic loads are considered in the design by either increased external loads or reduced resisting forces. Earthquakes, for example, are taken into account in the design of dams by summarising the dynamic loads and replacing them with an equivalent static load; DIN 19700-11 [49]. Axial cyclic loads on piles are considered in the design by a gradually decreasing wall friction and tip resistance with increasing number of cycles and amplitude; e.g. KORECK AND SCHWARZ [90]. However, most cyclic load cases have to be considered individual with respect to the planed construction and foundation, since neither regulations nor recommendations exist which propose a consistent, reliable and universally applicable calculation method.

1.1. Background

In frame of the Kyoto protocol, which aims to reduce lastingly the global CO_2 emission, many national governments have elaborated plans and promotional programmes to extend the amount of renewable energy on the total energy expenditure. Germany, for example, passed the *Erneuerbare-Energien-Gesetz* (EEG) [25] in 2004; a bill to support the development of new alternative energy sources and to regulate the feed-in remuneration. Other countries have passed similar bills.

It is expected that offshore wind energy will provide an important contribution to renewable energy. A study of the *German Wind Energy Institute* (DEWI) [52] estimates, that in 2030 15-20% of the German electricity consumption is covered by offshore wind.

However, by today, only some few considerable offshore wind parks are already in operation; e.g. *Horns Rev* [72] (Denmark, 160 MW, installed in 2002), *Nysted* [117] (Denmark, 158.4 MW, installed 2003) and *Prinses Amalia Windpark* [118] (Netherlands, 120 MW, installed 2008). All realised projects have in common, that they are founded near shore in shallow water in depths of maximum 20 m. The loads acting on these offshore wind turbines (OWT) are comparable to onshore installations. Wave loads and currents have only a minor influence on OWTs.

Larger projected offshore wind parks, from which some are already approved, are located outside the twelve nautical mile zone in areas with water depths of up to 40 m. While current OWTs have a performance of approximately 3 MW, future turbines will have a performance of at least 5 MW in order to be economical. These structures – and foundations – are significantly stronger exposed to cyclic loading due to wind, waves and currents. Additionally manufacturers of the turbines place high demands on the allowable permanent rotations of the substructure in order to guarantee a failure-free operation and low-maintainance.

Currently, five different foundation types are considered for OWTs, namely shallow foundations with and without skirts, monopiles and tri- or quadpods with piles, plates or buckets. Every offshore-founded structure has to be approved by national admission offices. In Germany the *Bundesamt für Seeschifffahrt und Hydrographie* (BSH) [29] is in charge of the certification. They demand that the design of the foundations has to consider besides the *Eurocode EC 7* [51] and the national standard *DIN 1054* [50] additional guidelines and regulations from *Germanischer Lloyd* [57], *Det Norske Veritas* [46] and the *American Petroleum Institute* [7, 8]. Although all standards, guidelines and regulations demand that the design has also to account for cyclic loading, none of them propose a reliable and directly applicable method.

The problems involved in the design of offshore structures are not new. Approaches to solutions have been proposed in the last decades by several authors; e.g. BJERRUM [24], EIDE AND ANDERSEN [55], RANDOLPH ET AL. [124] and ANDERSEN [11, 13] to name but a few. Their work was mainly focused on offshore facilities for the oil and gas industry. However, these structures differ significantly in several aspects from OWTs. Offshore wind turbines are relatively light structures, which are exposed to large imposed horizontal forces and overturning moments. Hence, even established methods from the oil and gas industry have to be reconsidered with respect to the special requirements of OWTs.

An important cyclic load case, which has been established in the design of offshore structures for the oil and gas industry, is the design storm. Thereto, all load reversals within a predefined time frame, generally 12 to 18 hours, are recorded and sorted in ascending order. That means, that the design storm starts with the smallest and ends with the largest load cycle. This is a reasonable assumption, since measurements have shown, that storms develop in a similar way. The intensity increases continuously and reaches its peak after three-fourth of the storm duration. Another important load case is the occurrence of a large single wave, or of a threesome of large waves, often denoted as *soliton* or *freak wave*; CLAUSS [40]. Both load cases are also relevant for OWTs, and should be adopted for their design. The dimensions of the forces and the load combinations have to be adjusted with respect to the site conditions and the shape of the structures.

In order to estimate the accumulated displacements and settlement of the structures during their lifetime, the translations and rotations within a design storm are summed up over the number of all expected storms in this time. All other smaller load reversals in between the storms are neglected. This is a conservative but also an unreliable approach. Due to this uncertainty, large and cost intensive foundations are most likely.

1.2. Scope of this work

This work aims to improve the design of foundations for OWTs by means of a more realistic approach. In particular the influence of smaller load cycles during regular operation will be considered. Main focus is spent on the accumulated rotations of the foundations, since they pose a particular challenge. Own model tests and numerical simulations should reveal if foundations are capable to back-rotate during small cyclic loading after an previous tilting due to a severe storm or a large single wave. Based on the rotational behaviour, stabilisation criterion should be derived.

The study focuses on shallow foundations, although other foundation types, such as monopiles, may behave similarly as model tests at the *Institute of Soil- and Rockmechanics* (IBF) have shown. Also, mainly drained conditions are considered, since the difference between fully and partially drained conditions is only qualitative with respect to the studied behaviour. The influence of pore pressure will be discussed, however, because it is relevant for a quantitative description of offshore foundations.

2. Foundation types and design for Offshore Wind Turbines

2.1. Locations and site conditions

The *Offshore Wind Competence Centre of the German Energy Agency* [47] provides on the front page of their Internet presence up-to-date information and interactive maps of the German continental shelf of the North and Baltic Seas, presenting the current state[1] of the projected offshore wind parks. They are all outside the twelve nautical mile (12-nm) zone located in water depths larger than 25 m. The operators plan to install turbines with capacities of at least 5 MW. The constructional works of the first offshore wind park in the German sector, *Alpha Ventus* [5], will start in spring 2009.

The subsoil consists mainly of Holocene medium dense to very dense sand deposits or Pleistocene stiff till and marl deposits. Due to the genesis the soil properties are varying within a wide range. Melting ice-age glaciers have formed flumes and water channels, which were back-filled with soft and fine-grained material. Large erratic boulders can also be encountered. A detailed summary is given by WIENMANN ET AL. [151].

2.2. Foundation types for offshore wind turbines

Figure 2.1 (RUBIN [135]) shows different possible foundation types for offshore wind turbines (OWT). Advantages and disadvantages with respect to cost-efficiency, construction management and further aspects are discussed in detail by e.g. ZAAIJER [155] and MUSIAL ET AL. [106]. This chapter focuses on geotechnical aspects only.

2.2.1. Gravity base structures

Gravity base structures (GBS), as shown in Figure 2.1a, are used successfully since several decades to found different kinds of offshore facilities for the oil and gas industry. They are generally made of concrete. 1973 the first GBS, the Ekofisk oil storage tank, was installed. It is located at the equally-named oil field in the North Sea; see e.g. CLAUSEN ET AL. [39]. The original tank had a circular shape with a diameter of approximately 90 m and a height of almost

[1] approval phase, already approved or in operation

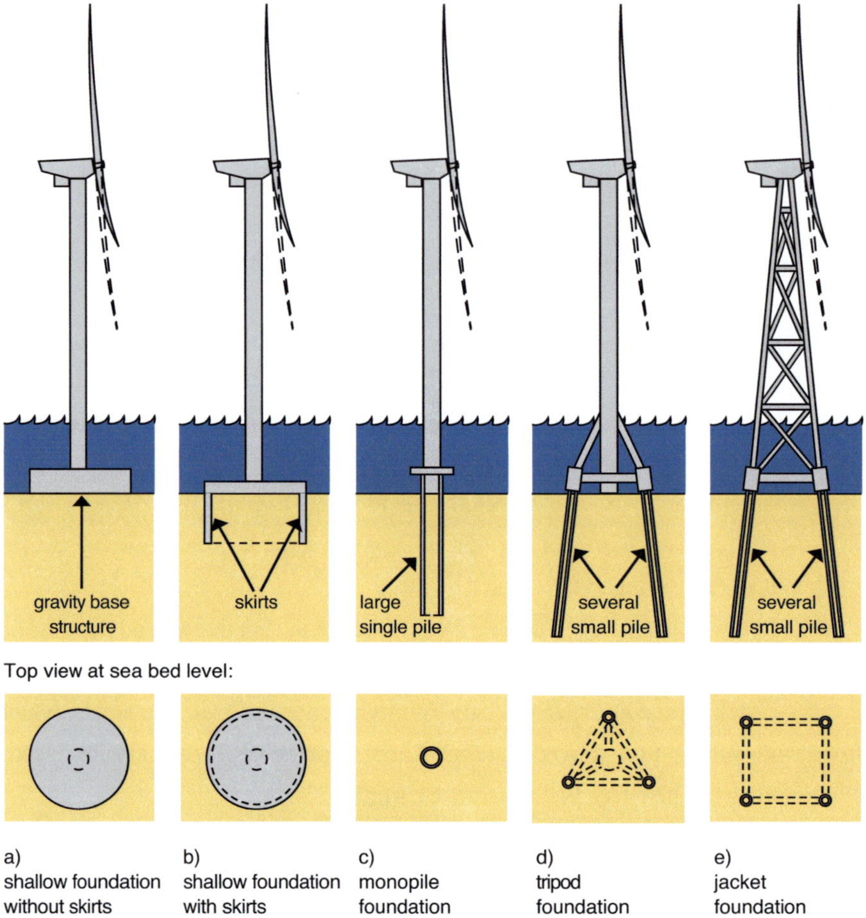

Figure 2.1.: Qualitative examples of foundation types for offshore wind turbines (RUBIN [135])

100 m. Due to critical amounts of accumulated settlement and subsidence, a protective barrier wall was placed in 1989 around the original tank. This wall should prevent the top side from getting flooded by large waves.

Although the settlement of installed GBSs were significant larger than original expected – values of up tp 25 cm per year were measured for the Ekofisk tank in the first years after installation – the overall good experiences made with GBSs yield to an increased confidence in this new offshore foundation type. Today, GBSs are employed successfully to install offshore structures on sand and clay. In 1995 the Troll A Condeep platform was installed at the Norwegian Troll field on a normal consolidated clay in a water depth of approximately 300 m; HANSEN ET AL. [65].

Today, all GBSs are equipped with relatively short skirts compared to the diameter of the foundations. These structural elements are made of concrete or steel and are attached on the bottom side of the foundations. They penetrate almost completely into the subsoil during installation. The skirts of the Ekofisk tank are made of concrete and have a length of 0.3 m. For comparison, the 180 m wide base of the Troll Condeep A platform has tubular steel skirts with a length of 32 m. Skirts have several functions. They

- serve as a protection against scour,
- increase the embedment depth of the foundation,
- can be used to apply additional vertical loads during installation by means of suction,
- and may serve as tool to readjust a foundation after a prohibitive tilting due to e.g. an accidental load.

The influence of skirts on the foundation behaviour has been intensively studied by e.g. LACASSE AND D'ORAZIO [91] and AAS AND ANDERSEN [1].

One of the few GBSs without skirts was the Frigg CDP 1 platform; LACASSE ET AL. [93] and LACASSE AND ROBBERSTAD [92]. Already shortly after the installation have divers discovered erosion and piping around the foundation base. The sand *breathed* which means that it flowed in and out the underside of the base in rhythm of the wave period. Furthermore, small cracks in the concrete diaphragm walls were observed, which indicated a temporary loss of contact between base and top sand in severe storms. Although, these problems and damages were not critical – according to the operators statement – the Frigg CDP 1 platform was disposed in 1990 after only 15 years of operation.

GBSs resist environmental forces by means of their weight and a large contact area. The effect of a possible embedment depth due to the presence of skirts is often neglected in the design. Horizontal forces acting on the superstructure are minimised by means of a suitable choice for the shape. A deep seated centre of gravity of the GBS and large diameter of the base help to resist large overturning moments.

Shallow foundations have some major advantages over deep foundations. They can be completely manufactured onshore. Thus, the required installation time on offshore site is short compared to deep foundations, which reduces the dependency on the weather conditions.[2] An-

[2] BJERRUM [24] reported, that bad weather conditions during constructions phase of several facilities at the Ekofisk field in the North Sea caused long hold-back times of expensive offshore equipment which are required to install deep foundations. This disadvantage accelerated the development of the first GBS.

other advantage of shallow foundations is the large diameter of the base. They can bridge local fluctuating soil properties.

GBSs have already been employed successfully to found OWTs; e.g. Nysted [117]. However, it is expected, that they are not uneconomical in deeper waters. Other than facilities for the oil and gas industry the overturning moments of OWTs are much larger. This would require very large base diameters. In order to employ GBSs in deeper water, the shape of the base has to be reconsidered. A possible solution will be presented in Chapter 5.

2.2.2. Bucket foundation

Figure 2.1b shows a *bucket foundation*, sometimes also denoted as *monopod*. It belongs to the group of shallow foundations. Unlike a GBS the bucket is a light-weight structure, but equipped with skirts of significant length. The ratio between foundation diameter d_0 and skirt length h_i is between 2.0 and 1.0 (see Figure 2.3). Substructures with $d_0/l_i \leq 0.5$ are called *suction anchors*. Latter are used to fix permanently floating offshore units, mainly in very deep waters; e.g. ANDERSEN ET AL. [17]. Buckets resist environmental forces by means of their weight, embedment depth and suction. In particular the ability to build up negative pore water pressure (suction) under severe loading conditions is an important feature of the bucket. Hence, they are mainly installed in soils with low permeability, such as fine dense sand or clay.

The skirts of bucket foundations are generally made of steel, while the base can be made of either concrete or also of steel. If the superstructure is founded on one bucket, as shown in Figure 2.1b, one speaks of *monopod*. Other common substructures consisting of buckets are the *tripod* and *quadpod*; that is a group of three, respectively, four buckets connected via a superstructure, which can be made of either steel or concrete. Two examples for offshore structures on buckets are the *Draupner E* and *Sleipner T* platforms; BYE ET AL. [32] and ERBRICH AND TJELTA [56]. Both are jacket-type superstructures, founded on four buckets (quadpod), each with a diameter of 12 m and skirt lengths of 6 m.

The bucket is a promising solution for founding OWTs, even in deeper waters. Currently, extensive model and field testing is done at the University of Oxford, UK, by HOULSBY ET AL. (e.g. [76]) and at the University of Aalborg, Denmark, by IBSEN ET AL. (e.g. [81]). Both focus on a monopod solution as shown in Figure 2.1b. The bearing behaviour of the monopod during loading is significantly different from that of a quadpod. A wave acting on the structures has two counteracting effects: it causes a large horizontal loading H and an asymmetrical increase of the vertical loading V; see Figure 2.2. The former yields at ground surface to an overturning moment $M(H)$ in the direction of the horizontal load while the latter yields to a moment $M(V)$ in opposite direction. The vertical load causes also a temporary increase of the submerged

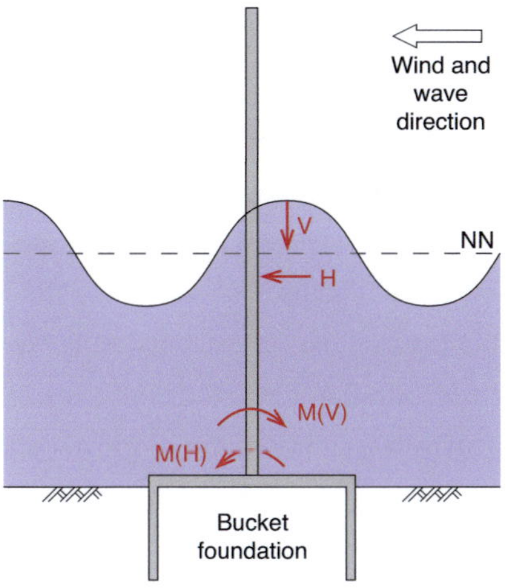

Figure 2.2.: Forces acting on a bucket foundation due to a large wave.

weight of the structure, which is an additionally restoring force. That means, the monopod resists overturning moments mainly by its rotational fixation in the subsoil.[3]

In order to gain a better understanding of the behaviour, a wind turbine with a bucket foundation was installed in 2003 in Frederikshaven, Denmark. Another test installation in Wilhelmshaven, however, failed. The penetration was stopped after the skirts started buckling. Probably the wall thickness and the stiffener were incorrectly designed.

However, a combination of a light-weight GBS and four buckets arranged to a quadpod, is, to the authors opinion, a very promising solution of OWTs. Some advantages of this foundation type are discussed in Chapter 5.

2.2.3. Monopile foundation

Figure 2.1c shows a monopile foundation. The monopile is a direct extension of the shaft of the superstructure through a transition piece into the subsoil. It consists of an one open-ended large steel pile which is completely penetrated into the seabed. The monopile resists environmental forces by horizontal bedding. By today, the monopile is the most often employed foundation type for OWTs.

An application of the monopile also in deeper waters is facing some major difficulties. The largest pile yet installed has a diameter of 5 m, a wall thickness of up approximately 100 mm and penetrates almost 40 m into the seabed. Installation of larger piles are currently not feasible with the available pile driving devices. Another difficulty is the, that the behaviour of monopiles

[3]in contrast to a GBS, which resists momentum loading by its weight only.

with diameters exceeding 5 m is not yet sufficiently understood and experience does not exist. Many past research activities focused on axially loaded piles (e.g. POULOS [119, 120, 121]), which is, however, at best relevant for pile groups. In order to account for horizontal loading, most recently proposed calculation procedures try to employ modified dynamic *p-y* and *t-z* curves. They are an extension to the procedures proposed by the *American Petroleum Institute* (API) [8]. Dynamic *p-y* and *t-z* curves should account for the soil's non-linearity and the energy dissipation through radiation damping, e.g. MOSTAFA AND NAGGAR [104]. Another, more promising approach, based on advanced numerical models, is followed by GRABE ET AL. [59] and by DÜHRKOP AND GRABE [144].

The monopile will be a serious solution for the foundation of OWTs, if the mentioned problems get solved. But the disadvantage of relatively long installation times of deep foundations over shallow foundations will persist.

2.2.4. Tripod foundation

In 1990, the first offshore wind turbine was installed in Nogersund, Sweden. It was founded on a tripod as shown in Figure 2.1d. This foundation type rests on three feet on the seabed and is commonly fixed with piles. Plates and buckets are also possible, although not yet accomplished. The tripod resists overturning moments by alternating push-pull load reversals of the piles.

Despite of the fact that the tripod was never employed again for OWTs, it is the favourite solution for German constructers and operators. The first German offshore wind turbines in the wind park Alpha Ventus will be founded on tripods.

The tripod is, to the authors opinion, a questionable solution. In particular a mechanism for a stabilisation behaviour, which is subject of this work, is lacking. There is no plausible explanation, which may cause a back-rotation after an initial tilting. The piles are subjected to alternate tension and compression loads during a severe storm. Hence they are successively pulled out. A re-penetration seems to be very unlikely.

Another disadvantage is the strong dependency on the weather conditions. One has to install three piles per turbine, which is generally done from a jack-up platform. This platform has to be, depending on the foundation geometry, replaced up to two times for each tripod.

2.2.5. Jacket foundation

The oldest foundation type employed for offshore structures is the jacket foundation. It is a steel construction fixed on the seabed by means of four piles or four pile groups. The concept originates from the oil and gas industry, for which it has been employed very often in different kind of soils and water depths.

The jacket foundation has the same disadvantages as the tripod. Hence, it is most likely an improper solution for the foundation of OWTs. Construction time and costs are expensive. Thus, jacket foundations are only interesting in deeper waters than in the currently projected offshore wind parks.

2.3. Design methods

The *Federal office for maritime navigation and hydrography* (BSH)[4] has to approve every in Germany constructed OWT before it can actually be built. Therefore the BSH has published a standard [29] in which they demand to provide evidence of safety against different kind of *failures*. The requirements are mainly based on the German standard DIN 1054 [50] and are extended for offshore-relevant design aspects, such as pore pressure generation or erosion.

The BSH standard reflects at best, to the authors opinion, the current state of knowledge in Germany relating design of foundations for offshore structures. For example, the BSH demands for gravity base structures to minimise the open gap between base and topsoil during an extreme loading. However, it is known since decades, that an open gap has to be avoided under offshore conditions in any case; e.g. BJERRUM [24]. since an open gap is accompanied by undermining and significant erosion.

It has been established in national and international standards and regulations to distinguish in the design of offshore strructures different so-called limit states: *ultimate limit state* (ULS), *serviceability limit state* (SLS), *fatigue limit state* (FLS) and *accidental limit state* (ALS). They should account for the safety against bearing failure and for the limitation of accumulated displacements of the foundations and accumulated strains of structural elements. However, the term *limit state* is misleading since arbitrarily predefined values of allowable translations/strains and rotations/torsions of a building/structural element are not a limit state. A limit state is characterised by a labile equilibrium, which responses to small changes of the boundary conditions with either a stabilisation or a sudden collapse. Due to this misconception, the terms ULS, SLS, FLS and ALS are rejected by the author and hence not used in this work.[5]

This section focuses on shallow foundations with skirts only, since deep foundations are not subject of this work. The available literature on comparisons of different supports structures for offshore wind turbines and the different design methods is confusing. A vast amount of papers and conference proceedings have been published in the last years. But only few authors

[4]Bundesamt für Seeschifffahrt und Hydrographie

[5]An exception is made in the appendix. There is used a load table which distinguish these different *limit states*.

examine critically the problems and challenges related with foundations for OWTs and propose reasonable solutions. Some will be presented in the following.

2.3.1. Installation

The first critical design issue is the installation. It has to be guaranteed that the skirts penetrate completely into the subsoil. Foundations with short skirts (GBS) penetrate by their weight only. Bucket foundations require additionally suction. This is achieved by means of pumps which pump out the water between bottom side of the plate and ground surface. In order to avoid damages of the skirts, in particular during the initial phase of the penetration, dowels are generally employed; e.g. MAZURKIEWICZ AND TOPOLNICKI [100], LACASSE AND D'ORAZIO [91] and AAS AND ANDERSEN [1]. They are attached on the bottom side of the foundation and are somewhat longer than the skirts. The dowels serve as a guidance and prevent critical horizontal movements of the foundation previous to the touch down of the skirt tips.

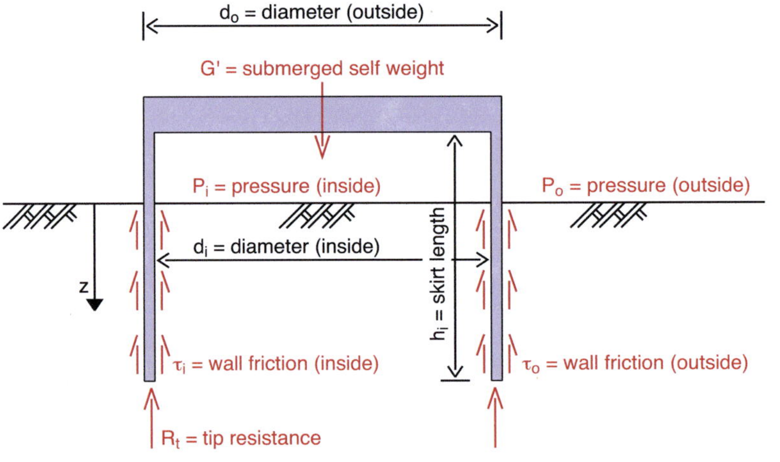

Figure 2.3.: Geometry and forces of a bucket foundation during penetration

The main point of interest is the achievable maximum penetration depth which depends on the soil properties (sand or clay, density, permeability, ...), maximum applicable suction and weight and shape of the foundation. Figure 2.3 shows the parameters relevant for the design. The penetration resistance force is the sum of the wall friction $T = A \cdot (\tau_i + \tau_a)$ and the tip resistance R_t. Both depend on the current penetration depth. The driving force is the sum of the submerged weight G' and the current suction pressure $P = P_i - P_0$. Both forces cannot be determined individually, since the suction influences due to seepage force also wall friction and tip resistance. The maximum penetration depth is reached, when the driving force gets equal to the resisting force.

Two different calculation procedures have been proposed in order to determine quantitative the different forces. Both were developed for bucket foundations and anchors.[6] The procedure proposed by HOULSBY AND BYRNE [73, 74] is an analytical approach, which bases on a bearing failure mechanism at the tip of the skirts. They consider different stress distributions at the tip, depending on the flow conditions and the penetration depth. Another rather empirical approach has been proposed by ANDERSEN AND JOSTAD [15] and ANDERSEN ET AL. [18]. They employ results of model tests and in-situ measurements on buckets and anchors in order to derive universally valid principles. Both procedures, however, make use of finite element simulations to determine the current seepage forces depending on the diameter of the foundation, the current penetration depth and the permeability of the soil. It is assumed, that the permeability of the soil is different inside and outside the bucket due to the installation process. These simulations have been done first by ERBRICH AND TJELTA [56] for the design of the *Draupner E* and *Sleipner T* platforms. Own calculations with the proposed models show, that both deliver similar results and can successfully predict in-situ tests.

2.3.2. In-service performance

In order to predict the in-service performance of offshore foundations, different load cases have to be considered. They are not independent, but may interact with each other.

It is known from numerous amount of laboratory tests, model tests and in-situ measurements, that dry sand samples and sand deposits subjected to alternate cyclic loading will densify; i.e. the pore volume (void ratio) will continuously decrease with more repetitions. If the pore volume is filled with water and the drainage is (partially) impeded, instead, pore water pressure will be built up. The rate[7] of pore pressure accumulation depends on the

- relative density of soil,
- initial shear stress ratio,
- cyclic shear stress ratio,
- existing pore pressure due to previous cycles.

[6] the installation is a minor issue for GBSs, since they are much heavier and have relatively shorter skirts

[7] derivative of a variable with respect to the number of cycles

This has been shown in cyclic laboratory tests on sand samples by e.g. BJERRUM [24], RAHMAN [123] and SEED [132]. Additionally, the accumulation of pore pressure depends in boundary value problems also on the

- storm characteristics (height, period and length of the waves),
- shape and dimensions of the structure,
- drainage conditions and compressibility of the soil.

In particular, previous load cycles may have an important effect. SEED [133], for example could show by means of field observations, that "*... deposits of sand subjected to low magnitude earthquakes, which are not sufficiently strong to cause liquefaction, will develop an increased resistance to liquefaction in subsequent earthquakes even though they may undergo no significant change in density.*".

An accumulation of pore water pressure is accompanied by a decrease of the effective mean stress. This yields to a gradual decrease of the incremental stiffness. That means, that average and cyclic displacements and settlement of a structure subjected to cyclic loading will increase.

Although the amount of achievable accumulated pore water pressure during a severe storm can be significant, the subsoil will not necessarily liquefy[8]. Especially very dense sand shows at the beginning of cyclic loading an increase of the pore pressure, but reaches asymptotically a stable state with $p^{av} \neq 0$; IBSEN [80]). This behaviour is described in the literature as cyclic mobility; e.g. CASAGRANDE [35], CASTRO [36, 37] and SEED [131]).

2.3.2.1. Static loading

An important load case, which has to be considered in the design of offshore structures, is the appearance of a large single load. This is comparable to a typical onshore loading condition. The engineer has to prove the safety against sliding, tilting and bearing failure. This can be done by means of either analytical or numerical calculations. Since the loading duration is generally shorter than the diffusion time governed by the dominating drainage path length and the permeability of the, it is reasonable to assume completely undrained conditions in the design.

Several authors have studied the failure behaviour of foundations with and without skirts for arbitrary combinations of vertical V, horizontal H and moment M loading, e.g. BYRNE AND HOULSBY [33, 34] and GOTTARDI ET AL. [58]. They showed that the analytical solutions proposed in national and international standards may underestimate the actual bearing capacity

[8] i.e. the effective stresses vanishes and the soil liquefies, unable to carry any shear stresses.

of foundations. The named authors could derive from model tests and numerical simulations a *cigar-shaped* failure surface in the V-H-$M/2R$ space; see Figure 2.4.

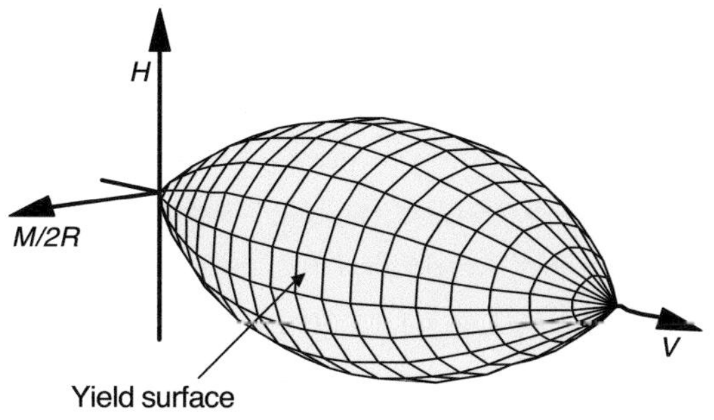

Figure 2.4.: Failure surface of a bucket derived from model tests (BYRNE AND HOULSBY [33])

The design becomes more complicated if the static load is applied on the structure subsequent to a number of smaller load cycles. They may affect the initial state of the soil. That means that the incremental stiffness of the subsoil can be either larger due to densification or smaller due to accumulated pore water pressure. A unique recommendation how cyclic loads should be taken into account does not exist, but has to be considered individually.

2.3.2.2. Cyclic loading

A key design issue is to predict the behaviour of a foundations subjected to cyclic loading; i.e. the amount and rate of densification and/or pore pressure accumulation. For this purpose mainly two approaches have been developed: a numerical and a (semi-) empirical one.

(a) Pore pressure contour diagram

(b) Shear strain contour diagram

Figure 2.5.: Results of undrained triaxial tests on Baskarp Cyclone Sand with relative density, Dr = 70% and 60% (JOSTAD et al. [86])

The empirical model bases on the concept of *equivalent number of cycles*; e.g. ALLOTEY AND EL NAGGAR [4]. The underlying idea is the hypothesis of a unique relationship between current effective mean pressure p, cyclic shear stress amplitude q^{ampl} and the number of cycles N. It says, that the number of cycles required to reach a certain amount of accumulated pore pressure decreases with increasing load amplitude. This is shown in Figure 2.5a. The curves represent isobars; i.e. lines with equal pore pressure. They are derived from a substantial number of undrained triaxial, biaxial or direct simple shear tests. Similar contour diagrams can be drawn for the accumulated shear deformation as shown in Figure 2.5b).

In order to use these diagrams to solve boundary value problems, the results of laboratory tests have to be prepared in machine-readable listings which can be used by e.g. a finite element program. The model discretises a considered subsoil by means of finite elements. During cyclic loading of the foundation the elements will be subjected to different cyclic loading conditions, which will be associated with the corresponding laboratory test, depending on the current stress and strain state. An alternative to the numerical simulation is an analytical approach. Thereto, reasonable failure mechanisms are assumed and the safety against failure is determined for each case.

The concept of equivalent number of cycles was derived first for clayey soils, since the influence of drainage can be neglected; e.g. ANDERSEN [12]. In order to use this concept in boundary value problems with seepage, i.e. with pore pressure accumulation and diffusion, two different extensions have been proposed. The first bases on an extensive laboratory testing program on samples with partially impeded drainage; NGI [107]. This procedure depends extremely on the considered boundary value problem. Thus only rather simple foundation geometries can be considered. Another, more reasonable approach is to employ again the finite element method; ANDERSEN ET AL. [14] and JOSTAD ET AL. [86]. It is an incremental and iterative procedure. The accumulation of pore pressure is determined for each cycles based on the same diagrams used for completely undrained conditions; e.g. Figure 2.5a. While the diffusion of pore water is computed by the finite element program based on Darcy's flow rule and an underlying simple constitutive model. The amount of diffusion depends on the geometry of the foundation and the permeability of the subsoil.

In any case, i.e. under drained or undrained conditions, an extensive laboratory testing is required, which is time consuming and expensive. For every new project and every foundation type and geometry further laboratory tests have to be performed.

An alternative to such (semi-) empirical models are numerical methods based on the finite element method and advanced constitutive models. One has to distinguish between implicit and explicit constitutive models. Implicit models trace numerically every load cycle, while explicit models consider average values only. The advantage of implicit models is that they can be used to simulate complex irregular loading conditions. But they have the disadvantage that they may accumulate a considerable numerical error within already a small number of cycles due to the employed predictor-corrector procedure. Explicit models are rather restricted to simpler loading paths, but can predict the accumulated stresses and strains for millions of load reversals.

The development of constitutive models has been accelerated in the last years due to a continuously increasing performance of computer systems. Besides the model used in this work, which has been proposed by NIEMUNIS ET AL. [113], presented in Chapter 3, there are some further explicit models worth mentioning: e.g. BOUCKOVALAS ET AL [27, 28] and more recently SAWICKI ET AL. [127, 128]. A comparison between the different models was presented by WICHTMANN [150].

2.3.3. Scour protection

A very important issue in the design of offshore foundations, is the protection against scour. One structural element which has been proved suitable for that, are skirts. They are, however, not sufficient for a successful protection against scour, but have to be used in conjunction with other protective measures. These are, e.g. with gravel or boulders whelmed geotextiles, which are attached to the foundation. Also bigbags besides the foundation filled with sand and other material have been proved suitable.

Scour protection is, despite of increasing advanced numerical simulations, still a mainly empirical job; DE GROOT ET AL. [45]. It always needs an extensive model test program; WHITEHOUSE [149].

2.4. Load assumptions and design loads

On an offshore wind turbine structure are acting several different regular loads such as wind, waves, currents and operational loads (from the turbine). They are difficult to quantify. Amplitude, frequency and directions are strongly fluctuating parameters; CLAUSS [40]. In order to prepare the loads for the design, several assumptions have to made, e.g. on the fundament type, geometry and dimensions. In general, wind and wave loads are determined independently. For the design of an OWT the relevant load combinations has to be derived. Two load combinations have been already mentioned before: the design storm, which should represent a severe storm, and a large single wave, a so-called soliton.

In the design of the structural elements of an OWT, e.g. blades, turbine or shaft, all load reversals are considered. Thereto, all expected load cycles are recorded and sorted in ascending order. The loads with the highest intensity are assumed to occur at the end of the projected life time of the OWT. Contrary to that, are considered in the design of the foundation and the soil-structure-interaction only some intensive load combinations, while intermediate smaller load cycles are neglected. Both approaches are unsatisfactory, since they are unrealistic. A reliable prediction of the behaviour of the foundation based on these assumptions is not possible.

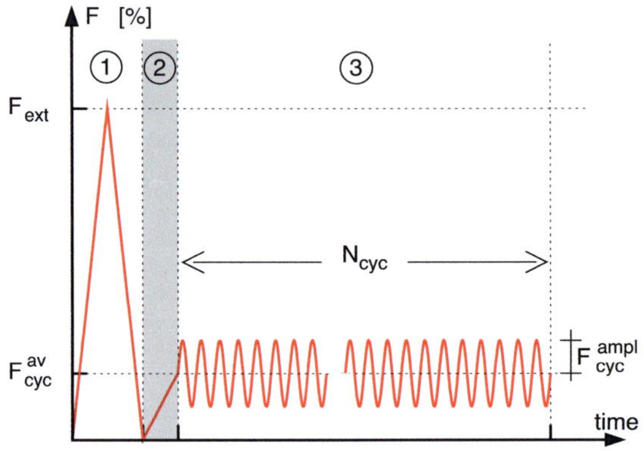

Figure 2.6.: Schematic representation of the employed loading sequence in model test and simulations. It consists of three load steps: ① extreme, ② intermediate and ③ cyclic loading step.

Since this work aims to improve the understanding, in particular of the influence of smaller load cycles during normal operation on the rotational behaviour of the foundation, a loading sequence as shown in Figure 2.6 has been employed in the model tests and the numerical study. The absolute values are taken from a load table derived from measurements at the projected offshore wind park *Borkum West*; Appendix C. It consists of three steps. The first one is an extreme load F_{ext} which represents a design storm or a soliton. It is applied once. In the second step the average load is applied. And in the third step the foundation is subjected to a series of N load reversals with constant amplitude F_{cyc}^{ampl} and average value F_{cyc}^{av}.

3. Constitutive models

Four different constitutive models were employed in order to study numerically the behaviour of shallow foundations subjected to cyclic loading. They will be presented in the following briefly. The basic mathematical formulations of the models are presented, but a detailed description of them is refrained from in this work in favour of qualitative and quantitative presentations of simulated laboratory element tests. Procedures for the determination of the model parameters are not discussed in detail, since no parameter sets have been determined in the frame of this work. Instead, references are given to other sources in the literature, in which the different models were originally presented.

3.1. Hypoplastic models and extensions

The term *hypoplasticity* was proposed first by WU [153]. It describes a group of constitutive models, which consist of a single tensorial equation. Contrary to elasto-plastic models, hypoplastic ones do not distinguish between elastic and plastic deformations nor between loading or unloading. Mathematical expressions for yield surface and flow rule, which are commonly employed in constitutive models, are not required.

Several hypoplastic formulations have been developed in the past. Worth mentioning is the CLoE model proposed by CHAMBON ET AL. [38] and the KARLSRUHE model proposed by KOLYMBAS [88]. Both models describe rate-independent behaviour of granular cohesionless materials such as sand or gravel. The in this work employed hypoplastic version of VON WOLFFERSDORFF [152] is a further development of the Karlsruhe model. Comparisons between different versions can be found e.g. in TAMAGNINI ET AL. [142] and LANIER ET AL. [95].

Apart from models for granular cohesionless materials further hypoplastic relations have been proposed for the description of viscous soils. Noteworthy are the models of NIEMUNIS [112, 109] and of GUDEHUS [61]. The former will be presented in the following.

3.1.1. Hypoplasticity

The hypoplastic relation by von Wolffersdorff [152] has the general form

$$\overset{\circ}{\mathbf{T}} = \mathsf{L} : \mathbf{D} + f_d \mathbf{N} \|\mathbf{D}\| \tag{3.1}$$

and describes the objective Zaremba-Jaumann stress rate $\overset{\circ}{\mathbf{T}} = \dot{\mathbf{T}} + \mathbf{T} \cdot \mathbf{W} - \mathbf{W} \cdot \mathbf{T}$ as a function of the current stress state $\mathbf{T}$, the stretching $\mathbf{D}$ and the void ratio e. The double brackets denotes the Euclidian norm of a tensor, defined as $\|\mathbf{D}\| = \sqrt{\mathrm{tr}\,\mathbf{D}^2} = \sqrt{\mathrm{tr}\,(\mathbf{D} \cdot \mathbf{D}^T)} = \sqrt{\mathbf{D} : \mathbf{D}}$. The fourth order tensor L and the second order tensor $\mathbf{N}$ are defined as:

$$\mathsf{L} = \frac{f_b f_e}{\hat{\mathbf{T}} : \hat{\mathbf{T}}} \left(F^2 \mathsf{I} + a^2 \hat{\mathbf{T}} \hat{\mathbf{T}} \right) = \frac{f_b f_e}{\hat{\mathbf{T}} : \hat{\mathbf{T}}} \hat{\mathsf{L}} \tag{3.2a}$$

$$\mathbf{N} = \frac{f_b f_e}{\hat{\mathbf{T}} : \hat{\mathbf{T}}} a F \left(\hat{\mathbf{T}} + \hat{\mathbf{T}}^* \right) = \frac{f_b f_e}{\hat{\mathbf{T}} : \hat{\mathbf{T}}} \hat{\mathbf{N}} \tag{3.2b}$$

with the normalised stress tensor $\hat{\mathbf{T}} = \mathbf{T}/\mathrm{tr}\,\mathbf{T}$ and the deviator $\mathbf{T}^* = \mathbf{T} - (1/3) \cdot \mathbf{1}\,\mathrm{tr}\,\mathbf{T}$. The tensors L and $\mathbf{N}$ incorporates the Matsuoka-Nakai failure condition [98] for $\overset{\circ}{\mathbf{T}} = \mathbf{0}$. The scalar factors f_d and f_e capture the dependency on the density (*pyknotropy*), while f_b captures the dependancy on the pressure (*barotropy*), viz.

$$f_d = \left(\frac{e - e_d}{e_c - e_d} \right)^{\alpha} \tag{3.3a}$$

$$f_b = \left(\frac{e_{i0}}{e_{c0}} \right)^{\beta} \frac{h_s}{n} \frac{(1 + e_i)}{e_i} \left(\frac{-\mathrm{tr}\,\mathbf{T}}{h_s} \right)^{(1-n)} \left[3 + a^2 - a\sqrt{3} \left(\frac{e_{i0} - e_{d0}}{e_{c0} - e_{d0}} \right)^{\alpha} \right]^{-1} \tag{3.3b}$$

$$f_e = \left(\frac{e_c}{e} \right)^{\beta} \tag{3.3c}$$

The exponents α and β are material constants. α determines the dilatancy behaviour under isobaric shearing, while β influences the incremental stiffness of dense materials under compression. The quantities e_{i0}, e_{c0} and e_{d0} are material constants, which represent characteristic void ratios at zero mean pressure ($p = 0$). The dependency of the limit void ratios e_i, e_c and e_d on the average pressure is described by Bauer's formula [20, 21].

$$\frac{e_i}{e_{i0}} = \frac{e_c}{e_{co}} = \frac{e_d}{e_{d0}} = \exp\left[-\left(\frac{-\mathrm{tr}\,\mathbf{T}}{h_s} \right)^{n} \right] \tag{3.4}$$

Therein, h_s and n are two further material parameters. h_s is called granular hardness. The exponent n ranges from about 0.2 to 0.6.

The failure condition obtained with the terms $\hat{\mathsf{L}}$ and $\hat{\mathbf{N}}$ for $\overset{\circ}{\mathbf{T}} = \mathbf{0}$ requires an additional material parameter, namely the critical friction angle φ_c.

$$a = \frac{\sqrt{3}(3-\sin\varphi_c)}{2\sqrt{2}\sin\varphi_c} \tag{3.5a}$$

$$F = \sqrt{\frac{1}{8}\tan^2\psi + \frac{2-\tan^2\psi}{2+\sqrt{2}\tan\psi\cos 3\theta}} - \frac{1}{2\sqrt{2}}\tan\psi \tag{3.5b}$$

$$\tan\psi = \sqrt{3}\|\hat{\mathbf{T}}^*\| \tag{3.5c}$$

$$\cos 3\theta = -\sqrt{6}\frac{\mathrm{tr}\,(\hat{\mathbf{T}}^*\cdot\hat{\mathbf{T}}^*\cdot\hat{\mathbf{T}}^*)}{(\hat{\mathbf{T}}^*:\hat{\mathbf{T}}^*)^{3/2}} \tag{3.5d}$$

The hypoplastic relation proposed by von Wolffersdorff requires eight constitutive parameters. They can be determined from standard laboratory tests and index tests. The procedures have been described in detail by HERLE [67] and HERLE AND GUDEHUS [68]. The model has been proved suitable to describe the stress-strain behaviour of loose and medium dense granular bodies subjected to monotonic loading and simple unloading under drained and undrained conditions; e.g. HERLE [67], CUDMANI [41] and NÜBEL [116].

This model has, however, some shortcomings for very dense materials. Figure 3.1 and Figure 3.2 show back-calculations of drained triaxial compression and extension tests; STURM [137]. They have been done in frame of a numerical simulation of centrifuge tests on a model-foundation of the Ekofisk oil-storage tank; ALLARD [2], ALLARD ET AL. [3] and ANDERSEN ET AL. [16]. The sand had an initial relative density of $D_r = 105\%$.

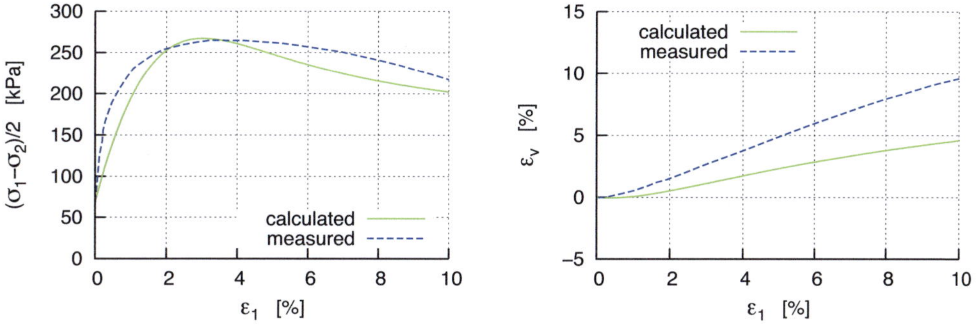

(a) Shear stress versus axial strain

(b) Volumetric strain versus axial strain

Figure 3.1.: Back-calculation [137] of a drained triaxial compression tests on very dense sand performed by the Norwegian Geotechnical Institute (NGI) [108]

While the predicted stress-strain curves agree well with measurements, Figure 3.1a and Figure 3.2a, the volumetric behaviour deviates significantly, Figure 3.1b and Figure 3.2b, since the

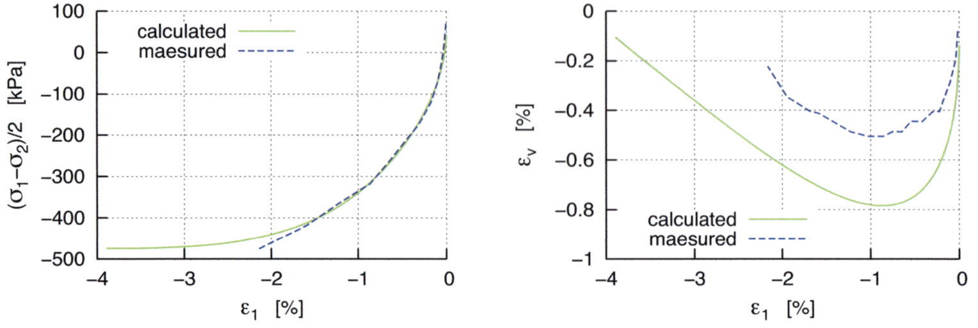

(a) Shear stress versus axial strain

(b) Volumetric strain versus axial strain

Figure 3.2.: Back-calculation [137] of a drained triaxial extension tests on very dense sand performed by the Norwegian Geotechnical Institute (NGI) [108]

Table 3.1.: Hypoplastic parameters used for the simulation shown in Figure 3.2.

h_s	n	e_{c0}	e_{d0}	e_{i0}	φ_c	α	β	m_T	m_R	R_{max}	β_χ	χ
625 MPa	0.33	1.05	0.67	1.21	32.8°	0.18	1.12	3.5	6.0	$1 \cdot 10^{-4}$	0.2	6

the computed dilatation is too low. NIEMUNIS [110] has shown, that the original hypoplastic relation of von Wolffersdorff, as used in the simulations shown in Figure 3.1 and Figure 3.2, can violate the condition $e \geq e_d$. Both shortcomings have been removed in version, which has been used in the numerical simulations presented in the subsequent sections. The pyknotropy factor f_d is replaced by

$$f_d = \operatorname{sign}(e-e_d)\left(\frac{e-e_d}{e_c-e_d}\right)^{\alpha} + \left[1-\operatorname{sign}(e-e_d)\left(\frac{|e-e_d|}{e_c-e_d}\right)^{\alpha}\right]^5 \bar{f}_d \tag{3.6}$$

$$\text{with} \quad \bar{f}_d = \frac{M_e^{(d)}\sqrt{3}(1+e) + M_T^{(d)} f_b f_e \frac{3}{\sqrt{3}}(3+a^2)}{M_T^{(d)} f_b f_e 3a},$$

$$M_e^{(d)} = 1\,, \quad M_T^{(d)} = -\frac{e_d}{h_s} n \left(\frac{-\operatorname{tr}\mathbf{T}}{h_s}\right)^{(n-1)}$$

which fortunately requires no additional material parameters. This modification prevents the current void ratio e from trespassing the lower bound e_d and improves the predicted volumetric strains of very dense granulates.

3.1.2. Visco-hypoplasticity

The visco-hypoplastic model proposed by NIEMUNIS [112, 109] aims to describe clayey and silty soils. In particular the model captures the rate-dependency of these materials. Niemunis rewrote Equation (3.1), by omitting the pyknotropy factors f_d and f_e, into

$$\overset{\circ}{\mathbf{T}} = f_b \hat{\mathsf{L}} : \left[\mathbf{D} - (-\hat{\mathsf{L}}^{-1} : \hat{\mathbf{N}}\,||\mathbf{D}||)\right] \tag{3.7}$$

and replaced the term $(-\hat{\mathsf{L}}^{-1} : \hat{\mathbf{N}}\,||\mathbf{D}||)$ by a viscous stretching $\mathbf{D}^{\text{vis}}$, viz.

$$\overset{\circ}{\mathbf{T}} = f_b \hat{\mathsf{L}} : \left[\mathbf{D} - \mathbf{D}^{\text{vis}}\right] \qquad \text{with} \qquad \mathbf{D}^{\text{vis}} = -D_r \vec{\mathbf{B}} \left(\frac{1}{\text{OCR}_{\text{B}}}\right)^{(1/I_v)} \tag{3.8}$$

Therein, D_r denotes a reference creep rate and I_v the viscosity index proposed by LEINENKUGEL [96]. Equation (3.8) resembles Norton's power law, which has been employed successfully to describe the creep behaviour of metals [115]. The hypoplastic flow rule tensor $\vec{\mathbf{B}}$ has been adopted in order to describe the direction of creep, viz.

$$\vec{\mathbf{B}} = \frac{\hat{\mathsf{L}}^{-1} : \hat{\mathbf{N}}}{||\hat{\mathsf{L}}^{-1} : \hat{\mathbf{N}}||} \tag{3.9}$$

The pyknotropy, i.e. the dependency of the material behaviour on the density, is allowed for only in the viscous stretching $\mathbf{D}^{\text{vis}}$, by means of an over-consolidation ratio

$$\text{OCR}_{\text{B}} = \frac{p_e}{p_e^+} \tag{3.10}$$

with a pressure p_e corresponding to the current void ratio e for the reference isotach. p_e is computed by means of the compression law proposed by BUTTERFIELD [31],

$$\ln\left[\frac{1+e_{e0}}{1+e}\right] = \lambda \ln\left[\frac{p_e}{p_{e0}}\right] \tag{3.11}$$

Therein, e_{e0} and p_{e0} are material parameters which define a reference point on the reference isotach. Actually, the reference isotach depends also on D_r, i.e. the three parameters e_{e0}, p_{e0} and D_r depend on each other. They can be neither defined independently nor determined uniquely. The compression index λ describes the inclination of the isotach, which is parallel but not necessarily identical to the virgin compression line. Equation (3.11) is also used in order to describe unloading and reloading with a swelling indey κ instead of λ.

The auxiliary pressure p_e^+ in Equation (3.10) is defined by

$$p_e^+ = \begin{cases} \frac{p}{\beta - 1}\left[\beta\sqrt{1+\bar{\eta}^2(\beta^2-1)} - 1\right] & \text{for} \quad \bar{\eta} < 1 \quad \text{(wet side)} \quad (3.12a) \\ p(1+\bar{\eta}^2)\frac{1+\beta}{2} & \text{for} \quad \bar{\eta} > 1 \quad \text{(dry side)} \quad (3.12b) \end{cases}$$

with a material parameter $\beta \approx 0{,}5 \div 0.95$ and $\bar{\eta}$ defined by

$$\bar{\eta} = \frac{q}{M\,p} \qquad \text{with} \qquad M = 6F(\mathbf{T})\frac{6\sin\varphi_c}{3-\sin\varphi_c}\,. \tag{3.13}$$

The function $F(\mathbf{T})$ is derived from the Matsuoka-Nakai failure condition as given in Equation (3.5b). Equation (3.12) projects the current stress state (p and q) onto the isotropic axis (p_e^+) by means of a modified cap-surface, taken from the modified Cam-Clay Model. Therein two different states are distinguished: contractant (Equation (3.12a)) – also denoted as *wet side* – and dilatant (Equation (3.12b)) – also denoted as *dry side*.

Finally, the barotropy factor f_b in Equation (3.8) has to be adjusted. Contrary to the hypoplastic model of von Wolffersdorff, f_b is to account for the influence of volume changes on the incremental stiffness in the absence of creep during unloading and reloading. By means of the time derivation of Equation (3.11), and κ instead of λ and $\mathbf{D}^{\text{vis}} = \mathbf{0}$, one obtains

$$f_b = -\frac{\operatorname{tr}\mathbf{T}}{\kappa\left(1+\frac{a^2}{3}\right)}\,. \tag{3.14}$$

The parameter a is a function of the critical friction angle φ_c, which arises from the Matsuoka-Nakai failure condition [98].

The visco-hypoplastic model by NIEMUNIS [112, 109] requires seven constitutive parameters. They can be determined by means of laboratory tests, preferably triaxial tests. But also oedometric, i.e. one-dimensional stress-controlled compression tests, can be used in order to determine the parameters, if the different stress state is considered.

Table 3.2.: Visco-hypoplastic parameters used for the simulation shown in Figure 3.3.

e_0	p_0	λ	κ	β	I_v [-]	$\dot{\gamma}$ [-]	φ_c	m_T	m_R	R_{max}	χ	β_χ
0.85	100 kPa	0.07	0.018	0.85	0.03	$1.0\cdot10^{-6}$	25°	10	10	$1\cdot10^{-4}$	0.058	1

Despite of its simplicity, the visco-hypoplastic model has been proved suitable to predict even complex loading paths. Figure 3.3 shows own back-calculation of one-dimensional compression tests on the so called Wiener Tegel clay performed by HVORSLEV [78]. The visco-

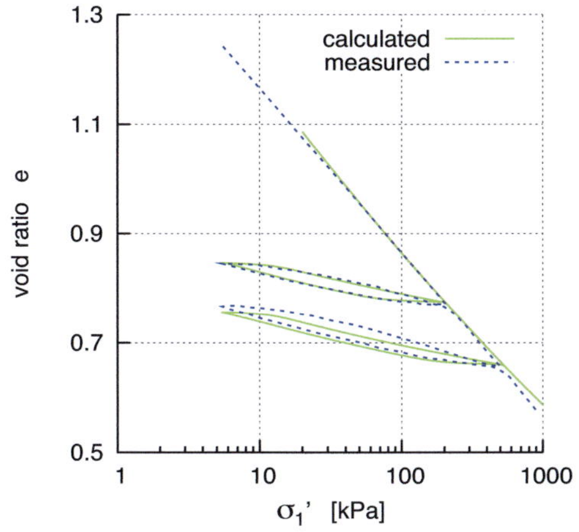

Figure 3.3.: Back-calculation of one-dimesnional compression tests done by HVORSLEV [78]

hypoplastic model can reproduce the compression behaviour as well as the two unloading-reloasing loops realistically. In order to describe the latter ones, the intergranular strain, presented in the next subsection, had to be employed.

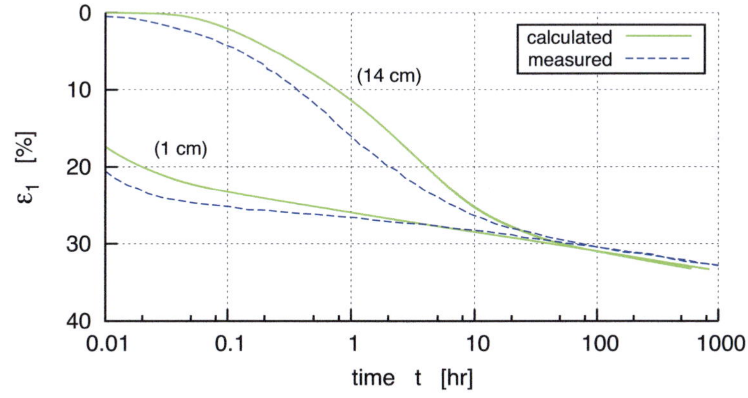

Figure 3.4.: Back-calculations of combined consolidation and creep tests on two samples with different initial heights performed by BARDEN [19])

Another important behaviour of viscous soils is the combined consolidation and creep behaviour under static total stress. The visco-hypoplastic model obeys the so-called hypothesis B, which says that the creep rate is independent of the sample height. This hypothesis has

Table 3.3.: Visco-hypoplastic parameters used for the simulation shown in Figure 3.4.

e_0	p_0	λ	κ	β	I_v [-]	$\dot{\gamma}$ [-]	φ_c	m_T	m_R	R_{max}	χ	β_χ
5.5	100 kPa	0.29	0.06	0.85	0.05	$1.0 \cdot 10^{-5}$	30°	5	5	$1 \cdot 10^{-4}$	0.1	1

been verified by e.g. IMAI AND TANG [82] and BARDEN [19].[1] Laboratory measurements of one-dimensional consolidation tests on two samples with different initial heights, performed by BARDEN [19], are presented in Figure 3.4. The back-calculation of these tests were done by means of the finite element method, because only the stress-strain-behaviour is computed by the constitutive model. The consolidation-behaviour should be computed on the discretised boundary value problem.

Table 3.4.: Visco-hypoplastic parameters used for the simulation shown in Figure 3.5.

e_0	p_0	λ	κ	β	I_v [-]	$\dot{\gamma}$ [-]	φ_c	m_T	m_R	R_{max}	χ	β_χ
0.7	100 kPa	0.1	0.01	0.85	0.047	$3 \cdot 10^{-7}$	42°	5	5	$1 \cdot 10^{-4}$	0.1	1

Figure 3.5 shows own back-calculations of undrained triaxial tests on Drammen clay, performed by BERRE AND BJERRUM [22]. In Figure 3.5a a comparison of measured and calculated results is presented for undrained shear stresses of clay samples which have initially the same state but are compressed with different axial strain rates. Figure 3.5b shows a derived result for the same tests. The normalised deviatoric stress decreases linearly – for given axial strain ε – with the logarithm of the strain rate $\dot{\varepsilon}$. One can see, that not only the stress-strain behaviour can be reproduced by the visco-hypoplastic model, but also the rate-dependence. The increase of the normalised shear stress is predicted qualitatively and quantitatively correct. The semi-logaritmic linear dependence, as shown in Figure 3.5b, is reproduced well. The deviation of the lower curve in the simulations can be explained by the employed parameter-triple D_r, e_{e0} and p_{e0}. The reference isotache should be slightly shifted, i.e. e_{e0} and D_r should be chosen somewhat larger, or p_{e0} somewhat lower for this curve.

Figure 3.6a to Figure 3.6d show own back-calculations [136] of biaxial tests on a remoulded saturated clay, performed by TOPOLNICKI [143]. The biaxial apparatus was a *Hambly-type* device with fixed vertical plates and two perpendicular movable pair of plates. Table 3.5 lists the different stretching conditions of the presented tests.

[1] Contrary to hypothesis B the so-called hypothesis A postulates, that both, the one-dimensional consolidation and the creep rate depend on the square of the sample height. This view is mainly persistently hold by MESRI AND CHOI [101, 102].

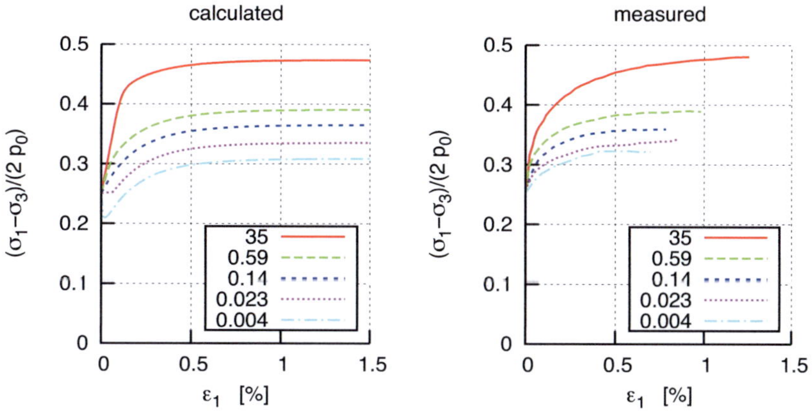

(a) Undrained triaxial compression tests with different strain rates

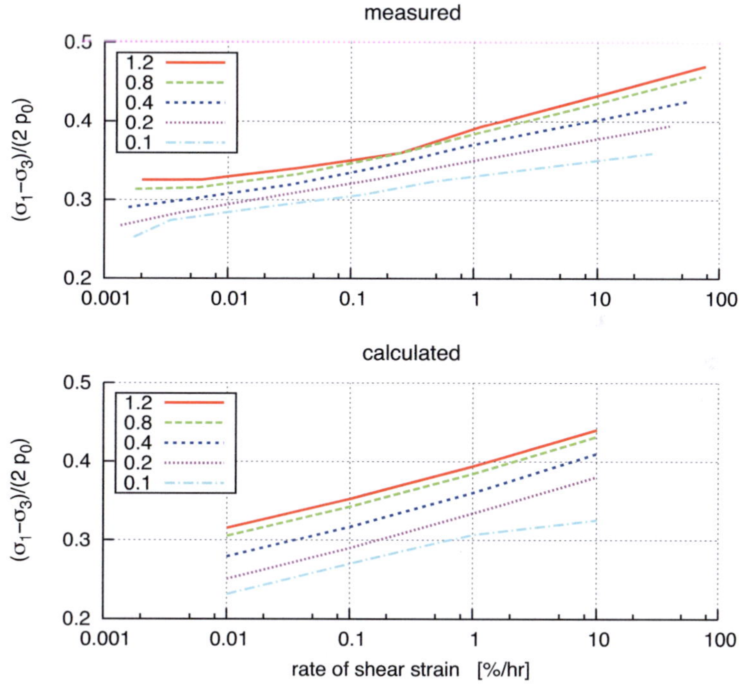

(b) Shear stress versus strain rate for a given axial strain

Figure 3.5.: Measurements by BERRE AND BJERRUM [22] and back-calculations of undrained triaxial tests on Drammen clay

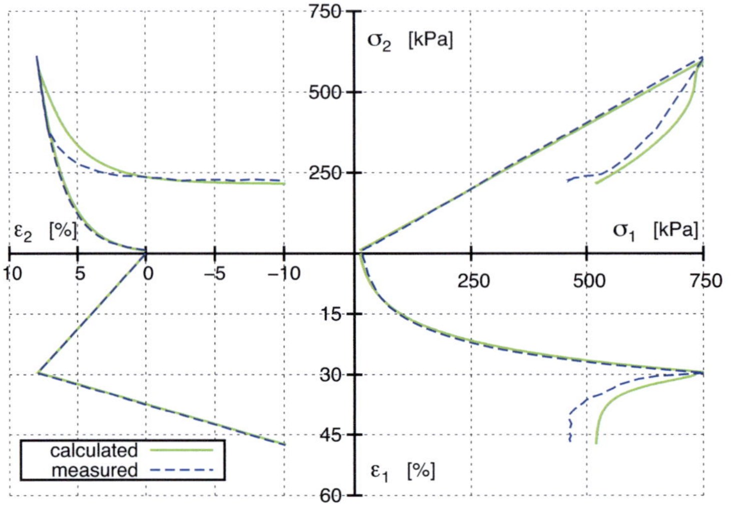

(a) Asymmetrical compression followed by an isochoric shearing

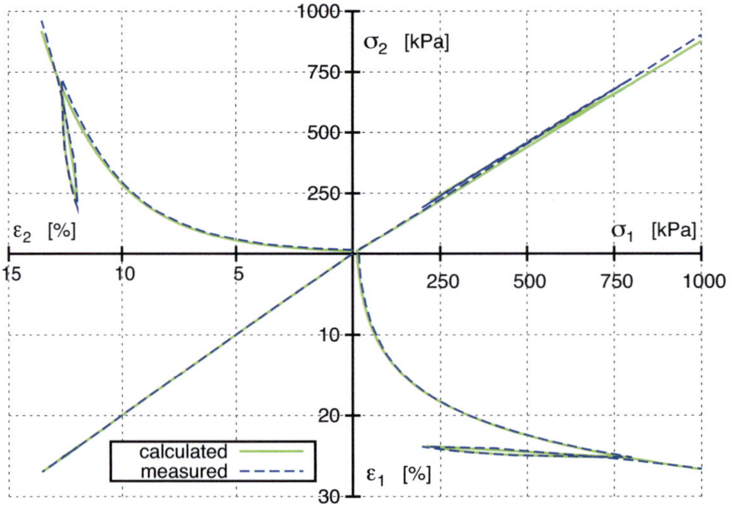

(b) Asymmetrical compression followed by an unloading and a re-compression

Figure 3.6.: Back-calculations [136] of biaxial tests performed by TOPOLNICKI [143]

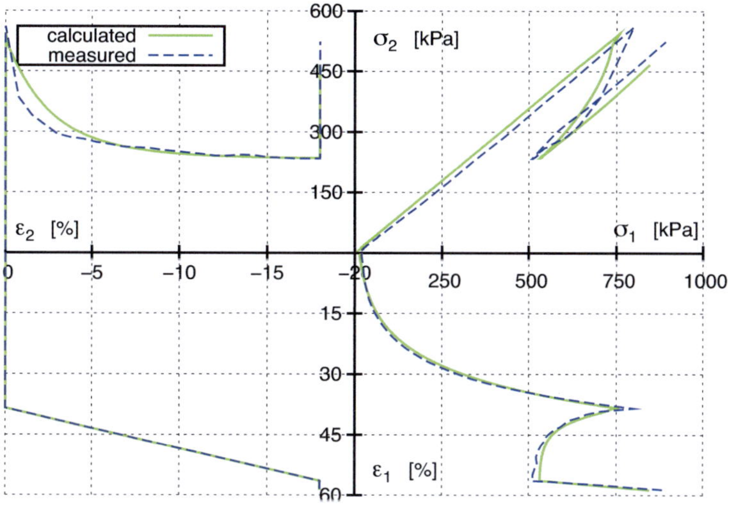

(c) Oedometric compression followed by an isochoric shearing and a further compression

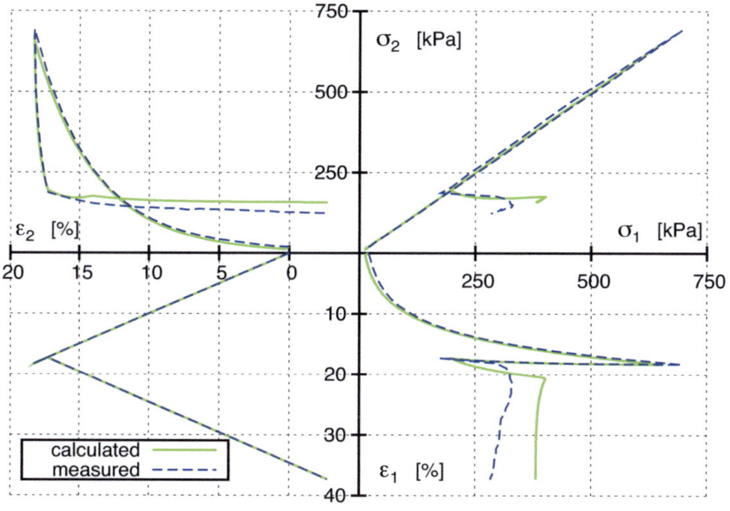

(d) Symmetrical (biaxial) compression followed by a symmetrical (biaxial) extension and an isochoric shearing

Figure 3.6.: Back-calculations [136] of biaxial tests performed by TOPOLNICKI [143]

Table 3.5.: Ratio of the logarithmic strain rates imposed by the plate pairs. Therein $\dot{\varepsilon}_1 = -\dot{\varepsilon}_2$ means isochoric shearing, $\dot{\varepsilon}_1 > -\dot{\varepsilon}_2$ compression and $\dot{\varepsilon}_1 < -\dot{\varepsilon}_2$ extension.

Figure	**Step 1**	**Step 2**	**Step 3**
Figure 3.6a	$\frac{\dot{\varepsilon}_1}{\dot{\varepsilon}_2} = \frac{4}{1}$	$\frac{\dot{\varepsilon}_1}{\dot{\varepsilon}_2} = -1$	-
Figure 3.6b	$\frac{\dot{\varepsilon}_1}{\dot{\varepsilon}_2} = \frac{2}{1}$	$\frac{\dot{\varepsilon}_1}{\dot{\varepsilon}_2} = -\frac{2}{1}$	$\frac{\dot{\varepsilon}_1}{\dot{\varepsilon}_2} = \frac{2}{1}$
Figure 3.6c	$\frac{\dot{\varepsilon}_1}{\dot{\varepsilon}_2} = \frac{1}{0}$	$\frac{\dot{\varepsilon}_1}{\dot{\varepsilon}_2} = -1$	$\frac{\dot{\varepsilon}_1}{\dot{\varepsilon}_2} = \frac{1}{0}$
Figure 3.6d	$\frac{\dot{\varepsilon}_1}{\dot{\varepsilon}_2} = \frac{1}{1}$	$\frac{\dot{\varepsilon}_1}{\dot{\varepsilon}_2} = -\frac{1}{1}$	$\frac{\dot{\varepsilon}_1}{\dot{\varepsilon}_2} = -1$

Table 3.6.: Visco-hypoplastic parameters used for the simulation shown in Figure 3.6.

e_0	p_0	λ	κ	β	I_v [-]	$\dot{\gamma}$ [-]	φ_c	m_T	m_R	R_{max}	χ	β_χ
0.96	100 kPa	0.09	0.025	0.9	0.031	$1.3 \cdot 10^{-5}$	21.5°	5	5	$1 \cdot 10^{-4}$	0.04	1

One can see, that a remarkable agreement between simulations and measurements is obtained. In all simulations the same parameter set has been employed. The calibration has been done on another set of laboratory tests, which are not shown here; STURM [136]. Hence, the simulations shown in Figure 3.6a to Figure 3.6d are rather a predictions than a back-calculations. The good agreement could only be achieved after considering short resting periods between the different loading steps, as listed in Table 3.5. After installation and pre-consolidation, the samples in the tests got slightly over-consolidated due to circuit times of up to several minutes. Hence, the initial $\mathrm{OCR}_{\mathrm{B,\,ini}}$ rises from 1 up to 1.1 due to relaxation.

The last example, shown in Figure 3.7, presents laboratory measurements and simulations of the development of creep rate and creep failure. The measurements (Figure 3.7a) were performed with drained triaxial tests; MURAYAMA AND SHIBATA [23, 105]. All samples had the same initial state with $\mathrm{OCR}_{\mathrm{B}} > 1.0$. They were then subjected to different stress ratios $\eta = q/p$, which was maintained in the sequel. Normally, the creep rate decreased with axial strain and hence time. However, if the applied shear stress $\eta \cdot p = q$ was close to the maximum possible shear stress q_{max}, an acceleration of the axial strain rate after a deceleration could be observed; top curve in Figure 3.7a. This behaviour is called creep failure. During the creep phase, the sample dilates which is accompanied by a decrease of shear strength.

A simulation of this behaviour is shown in Figure 3.7b. Due to numerical difficulties of stress-controlled tests, drained direct simple shear tests were simulated, instead of the more

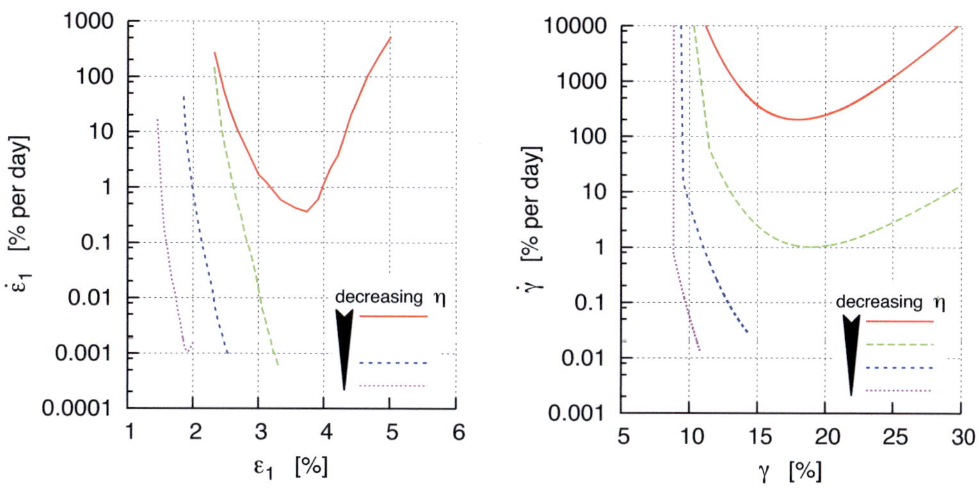

(a) Measured creep tests in drained triaxial tests performed by MURAYAMA AND SHIBATA [105]

(b) Calculated creep tests in a direct simple shear test

Figure 3.7.: Measured and calculated creep behaviour and creep-failure in laboratory tests at constant stress ratios $\eta = q/p$

Table 3.7.: Visco-hypoplastic parameters used for the simulation shown in Figure 3.7.

e_0	p_0	λ	κ	β	I_v [-]	$\dot{\gamma}$ [-]	φ_c	m_T	m_R	R_{max}	χ	β_χ
0.7	100 kPa	0.083	0.026	0.85	0.025	$1 \cdot 10^{-6}$	18.5°	3	3	$1 \cdot 10^{-4}$	0.07	1

complicated triaxial tests. Both, the initial decrease of the shear strain rate, as well as the delayed acceleration in case of higher stress ratios can be predicted by the visco-hypoplastic constitutive model. The results agrees qualitatively well with the observations.

3.1.3. Intergranular strain

The intergranular strain is an extension to the hypoplastic model of von Wolfferdorff and the visco-hypoplastic model of Niemunis. The necessity of this extension was, however, different for both models, which will be discussed in the following subsections.

3.1.3.1. Hypoplasticity with intergranular strain

The hypoplastic model has been proved suitable for monotonous loading and at best one unloading, but it has shortcomings for cyclic loading. The incremental stiffness during reloading after unloading is under-estimated. The hypoplastic differential stiffness for reloading is almost equal to the one for the first loading.[2] This shortcoming produces an exaggerated *ratcheting*; shown in Figure 3.8. Ratcheting denotes here the accumulation of permanent deformations, stresses or both, depending on the loading conditions, during cyclic loading. In order to improve the hypoplastic model, NIEMUNIS AND HERLE [111] introduced the concept of *intergranular strain*. They replaced the tensors L and $\mathbf{N}$ in the basic hypoplastic formulation (Equation (3.1))

$$\overset{\circ}{\mathbf{T}} = \mathsf{L} : \mathbf{D} + f_d \mathbf{N} \|\mathbf{D}\| \quad \Rightarrow \quad \overset{\circ}{\mathbf{T}} = \mathsf{M} : \mathbf{D} \tag{3.15a}$$

by means of a new incremental stiffness tensor M and a switch function, which distinguishes between primary loading and reloading, viz.

$$\mathsf{M} = [\rho^{\chi} m_T + (1-\rho^{\chi}) m_R] \mathsf{L} + \begin{cases} \rho^{\chi}(1-m_T)\mathsf{L} : \vec{\mathbf{h}}\vec{\mathbf{h}} + \rho^{\chi} f_d \mathbf{N}\vec{\mathbf{h}} & \text{for} \quad \vec{\mathbf{h}} : \mathbf{D} > 0 \\ \rho^{\chi}(m_R - m_T)\mathsf{L} : \vec{\mathbf{h}}\vec{\mathbf{h}} & \text{for} \quad \vec{\mathbf{h}} : \mathbf{D} \leq 0 \end{cases} \tag{3.15b}$$

and $\rho = \|\mathbf{h}\|/R$, with R being the amount of the intergranular strain and $\vec{\mathbf{h}}$ the current direction.[3] The components of $\vec{\mathbf{h}}$ represent the recent deformation direction. After a strain path reversal, the intergranular strain tensor swivels towards the current deformation direction. The evolution equation for $\mathbf{h}$ is

$$\overset{\circ}{\mathbf{h}} = \begin{cases} (\mathbf{1} - \vec{\mathbf{h}}\vec{\mathbf{h}}\rho^{\beta_R}) : \mathbf{D} & \text{for} \quad \vec{\mathbf{h}} : \mathbf{D} > 0 \\ \mathbf{D} & \text{for} \quad \vec{\mathbf{h}} : \mathbf{D} \leq 0 \end{cases} \tag{3.15c}$$

with the same switch function. The incremental stiffness is increased with respect to the original one, depending on the amount of relative rotation, and becomes identical with the original one if $\mathbf{h}$ and $\mathbf{D}$ have the same direction.

This extension improves significantly the predicted behaviour of cyclically loaded granular materials. A comparison between original and extended model is shown in Figure 3.8. It shows the development of accumulated pore-pressure in an undrained triaxial test with symmetrical deviatoric load cycles and constant amplitude. The original hypoplastic model reaches within some few cycles a periodical state characterised by a double stress loop (the so called butterfly). Contrary to that, the model with intergranular strain requires a considerable amount of load reversals until it reaches the same state, which agrees well with laboratory measurements, performed by e.g. WICHTMANN [150] and HYODO ET AL. [79].[4]

[2] The hypoplastic differential stiffness during reloading is actually different from the one for the first loading, since the void ratio is different

[3] $\vec{\sqcup}$ denotes the direction of a variable viz. $\vec{\sqcup} = \sqcup/\|\sqcup\|$

[4] Most tests presented in the literature were performed with drained traixial tests. For this case, plastic deformations instead of pore pressure accumulates.

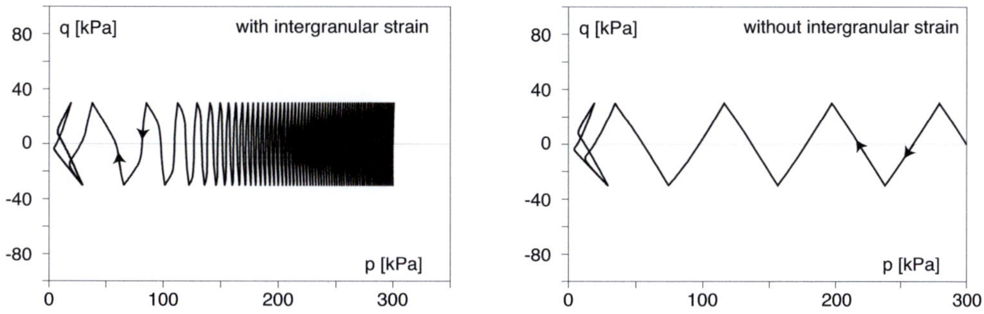

Figure 3.8.: Comparison between hypoplastic model with and without intergranular strain (NIEMUNIS [111])

The intergranular strain requires five additional parameters, namely m_T, m_R, R, χ and β_χ. Unfortunately, only a crude method for the determination has been suggested by NIEMUNIS [110] and NIEMUNIS AND HERLE [111]. The parameters can be estimated from results of resonant column tests and high-resultion triaxial tests with rather complicated strain paths. The difficulty is, that the intergranular strain parameters interfere with some hypoplastic parameters. In particular the two exponents α and β are affected. Hence, the parameters of the extended hypoplastic model have to be matched by trial and error.

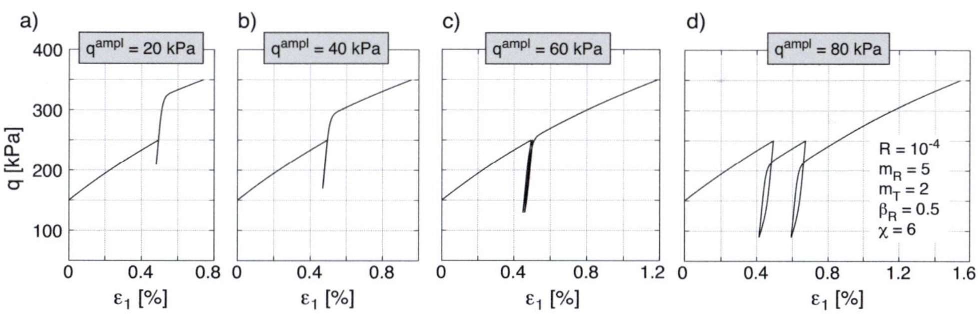

Figure 3.9.: Simulations of a triaxial test with two interposed load cycles (WICHTMANN [150])

A major disadvantage of the intergranular strain is, that its parameters are state-dependent, as shown in Figure 3.9. WICHTMANN [150] performed some numerical simulations of drained triaxial compression tests with two interposed load cycles. He increased in every test the amplitude of the cycles. In all simulations the same parameter set was employed. For small stress amplitudes, $q^{\text{ampl}} = 20$ kPa and $q^{\text{ampl}} = 40$ kPa, the predicted shear stress subsequent to the cycles is too large. Contrary to that, the predicted shear stress is too small for large cyclic amplitudes, $q^{\mathrm{ampl}} = 80$ kPa). A realistic prediction is obtained only for $q^{\mathrm{ampl}} = 60$ kPa, which is the stress range for which the parameters listed in Figure 3.9d) have been determined.

The problem is, that the intergranular strain is defined in the strain space and is assumed to be independent of the current stress state and density. Hence, one has to keep in mind during the evaluation of simulations of boundary value problems that the stresses should be in the same range as the ones assumed for the determination of the parameter set.

3.1.3.2. Visco-hypoplasticity with intergranular strain

NIEMUNIS [110] has originally proposed the concept of an intergranular strain for a visco-hypoplastic model in order to describe the hysteretic behaviour during unloading and reloading loops. Other than in the hypoplastic model, however, the intergranular strain is not needed to compensate for an excessive ratcheting. The visco-hypoplastic model yields already considers different incremental stiffnesses for primary loading and for reloading by means of the two parameters λ and κ.

For reasons of numerical stability, only the linear fourth-order tensor $\hat{\mathsf{L}}$ is replaced by a stiffness tensor $\hat{\mathsf{M}}$, which depends on the intergranular strain, viz.

$$\overset{\circ}{\mathbf{T}} = f_b \left[\hat{\mathsf{L}} : \mathbf{D} - \hat{\mathsf{L}} : \mathbf{D}^{\mathrm{vis}}\right] \qquad \Rightarrow \qquad \overset{\circ}{\mathbf{T}} = f_b \left[\hat{\mathsf{M}} : \mathbf{D} - \hat{\mathsf{L}} : \mathbf{D}^{\mathrm{vis}}\right] \tag{3.16a}$$

with

$$\hat{\mathsf{M}} = [\rho^{\chi} m_T + (1-\rho^{\chi}) m_R]\hat{\mathsf{L}} + \begin{cases} \rho^{\chi}(1-m_T)\hat{\mathsf{L}} : \vec{\mathbf{h}}\vec{\mathbf{h}} & \text{for} \quad \vec{\mathbf{h}} : \mathbf{D} > 0 \\ \rho^{\chi}(m_R - m_T)\hat{\mathsf{L}} : \vec{\mathbf{h}}\vec{\mathbf{h}} & \text{for} \quad \vec{\mathbf{h}} : \mathbf{D} \leq 0 \end{cases} \tag{3.16b}$$

Equation (3.15c) is also taken for the extension to the visco-hypoplastic model. Meaning and determination of the parameters are likewise similar. An interference between intergranular strain parameter and visco-hypoplastic parameters was not observed.

Figure 3.3 and Figure 3.6b show the influence of the intergranular strain during unloading and reloading loops. Without the extension, both paths would have fallen together on one line with an inclination of $\approx \kappa$.

3.2. SaniSand

SaniSand, which has been developed by DAFALIAS AND MANZARI [43] and recently improved by TAIEBAT AND DAFALIAS [141], belongs to the group of elasto-plastic constitutive models.

I.e. it distinguishes explicitly between loading (Equation (3.17a)) and unloading (Equation (3.17b)), respectively, (hypo-)elastic and elasto-plastic deformations.

Starting with the basic elasto-plastic formulation and non-associated a flow rule

$$\overset{\circ}{\mathbf{T}} = \begin{cases} \left[\mathsf{E}^{el} - \dfrac{\mathsf{E}^{el} : \mathbf{MN} : \mathsf{E}^{el}}{K + \mathbf{N} : \mathsf{E}^{el} : \mathbf{M}}\right] : \mathbf{D} & \text{for loading } (\mathbf{n} : \mathsf{E}^{el} : \mathbf{D} > 0) \quad (3.17a) \\ \mathsf{E}^{el} : \mathbf{D} & \text{for unloading } (\mathbf{n} : \mathsf{E}^{el} : \mathbf{D} < 0) \quad (3.17b) \end{cases}$$

with $\mathbf{M}$ being the flow rule, $\mathbf{N}$ the flow direction, K the plastic modulus and E^{el} the tangential isotropic elastic stiffness by

$$\mathsf{E}^{el} = 2G^{el}\mathsf{I} + \left(K^{el} - \frac{2G^{el}}{3}\right)\mathbf{1}\mathbf{1}\,. \tag{3.18a}$$

Therein, I denotes the identity tensor with $\mathsf{I} = \frac{1}{2}(\delta_{ik}\delta_{jl} + \delta_{il}\delta_{jk})$ and G^{el} the tangential elastic shear bulk modulus. This means

$$K^{el} = \frac{2}{3}\frac{(1+\nu)}{(1-2\nu)}G^{el} \tag{3.18b}$$

$$G^{el} = G_0^{el}\, p_{atm}\frac{(2.97-e)^2}{1+e}\sqrt{\frac{p}{p_{atm}}} \tag{3.18c}$$

The variables ν, G_0^{el} are elastic material parameters, and p_{atm} is a reference pressure. e denotes the current void ratio and p the mean pressure ($p = -\frac{1}{3}\mathrm{tr}\,\mathbf{T}$). The yield surface of the elasto-plastic model is given by

$$f = p\left[\|\mathbf{r} - \alpha\| - m\sqrt{\frac{2}{3}}\right] \overset{!}{=} 0 \tag{3.19a}$$

with the stress obliquity $\mathbf{r} = \mathbf{T}^*/p$ and a further material parameter m describing the opening angle of the yield surface. The tensor α is a hidden state variable which may be considered as a deviatoric back-stress ratio. The derivative of the non-associative flow rule f with respect to $\mathbf{T}$ is

$$\frac{\partial f}{\partial \mathbf{T}} = \frac{1}{\|\mathbf{r} - \alpha\|}\left[(\mathbf{r} - \alpha) + \frac{1}{3}(\mathbf{r} - \alpha) : \mathbf{r}\,\mathbf{1}\right] \tag{3.19b}$$

which is required to compute the so-called loading direction

$$\mathbf{N} = \frac{\dfrac{\partial f}{\partial \mathbf{T}}}{\|\dfrac{\partial f}{\partial \mathbf{T}}\|} \tag{3.20}$$

DAFALIAS AND MANZARI [43] proposed a non-associative flow rule $\mathbf{M} \neq \mathbf{N}$. Because of numerical instabilities in case of small load reversals, a modified flow rule $\mathbf{M}_{\text{modified}}$ was proposed by Niemunis abd PRADA [122] in a recent version of SaniSand model, which was used for the presented study, viz.

$$\mathbf{M}_{\text{original}} = \frac{B\mathbf{n} - C\left(\mathbf{n}\cdot\mathbf{n} - \frac{1}{3}\mathbf{1}\right) + \frac{1}{3}D\mathbf{1}}{\|B\mathbf{n} - C\left(\mathbf{n}\cdot\mathbf{n} - \frac{1}{3}\mathbf{1}\right) + \frac{1}{3}D\mathbf{1}\|} \tag{3.21a}$$

$$\mathbf{M}_{\text{modified}} = \frac{B\mathbf{n} + \frac{1}{3}D\mathbf{1}}{\|B\mathbf{n} + \frac{1}{3}D\mathbf{1}\|} \tag{3.21b}$$

with

$$B = 1 + \frac{3}{2}\frac{g\,(1-c)}{c}\cos(3\,\theta_{\mathbf{n}}) \tag{3.22a}$$

$$\text{with} \quad g = \frac{2c}{(1+c) - [(1-c)\cos(3\,\theta_{\mathbf{n}})]}, \qquad c = \frac{3-\sin\varphi_c}{3+\sin\varphi_c},$$

$$\cos(3\,\theta_{\mathbf{n}}) = -\sqrt{6}[(\mathbf{n}\cdot\mathbf{n}):\mathbf{n}] \quad \text{and} \quad \mathbf{n} = \frac{\mathbf{r}-\alpha}{\|\mathbf{r}-\alpha\|}$$

$$C = 3\sqrt{\frac{3}{2}}\frac{(1-c)}{c}g \tag{3.22b}$$

$$D = -A_d(\alpha_\theta^d - \alpha):\mathbf{n} \quad \text{with} \quad A_d = A_0\left[1 + \left(\frac{\mathbf{z}:\mathbf{n} + |\mathbf{z}:\mathbf{n}|}{2}\right)\right] \tag{3.22c}$$

Therein D is a dilatancy coefficient, which is a function of the current inclination α and the so-called accumulated fabric dilatancy deviator $\mathbf{z}$. The factor C, which may cause a numerical instability under certain loading conditions, has been omitted in M_{modified} (Equation (3.21b)). A_0 and φ_c are material parameters.

The plastic modulus is defined as

$$K = -\left[\frac{1}{\|f'\|}f' : \frac{\partial\dot{\alpha}}{\partial\dot{\varepsilon}^{\text{pl}}}\right] : \mathbf{M} \tag{3.23}$$

with $\dot{\varepsilon}^{\text{pl}} = \mathbf{M}\dot{\lambda}$. The plastic multiplier $\dot{\lambda}$ has the form

$$\dot{\lambda} = \frac{\mathbf{N} : \mathsf{E}^{\text{el}} : \mathbf{D}}{\mathbf{N} : \mathsf{E}^{\text{el}} : \mathbf{M} + K} \tag{3.24a}$$

and the evolution of the deviatoric back-stress $\dot{\alpha}$ is described by

$$\dot{\alpha} = \left(\frac{\dot{\lambda} + |\dot{\lambda}|}{2}\right) \frac{2}{3} h \left(\alpha_\theta^b - \alpha\right) \tag{3.24b}$$

with

$$h = \frac{G_0 h_0 (1 - c_h e)}{\sqrt{\frac{p}{p_{\text{atm}}} (\alpha - \alpha_{\text{ini}})}} . \tag{3.24c}$$

Therein h_0 and c_h are two further material parameters. The record variable α_{ini} is used in order to store the point of the most recent stress reversal outside the elastic range.

The evolution of the accumulated fabric dilatancy deviator $\mathbf{z}$ is described by

$$\dot{\mathbf{z}} = -c_z \left(\frac{\dot{\varepsilon}_v^{\text{pl}} + |\dot{\varepsilon}_v^{\text{pl}}|}{2}\right) (z_{\max} \mathbf{n} + \mathbf{z}) \tag{3.25}$$

with two material parameters c_z and $z_{\max}$. The plastic volumetric strain rate is $\dot{\varepsilon}_v^{\text{pl}} = \dot{\lambda} \operatorname{tr} \mathbf{M}$.

The SaniSand version of Dafalias and Manzari [43] has two yield surfaces, α_θ^b and α_θ^d, the version of Taiebat and Dafalias [141] a third one α_θ^c:

- an inboard dilatancy surface (also called phase transition line)

$$\alpha_\theta^d = \mathbf{n} \sqrt{\frac{2}{3}} [g M^c \exp(n^d [e - e_c]) - m] \tag{3.26a}$$

- an outboard bounding surface

$$\alpha_\theta^b = \mathbf{n} \sqrt{\frac{2}{3}} [g M^c \exp(-n^b [e - e_c]) - m] \tag{3.26b}$$

$$e_c = e_{c0} - \lambda \left(\frac{p}{p_{\text{atm}}}\right)^\xi \tag{3.26c}$$

- and an intermediate critical surface (also called critical state line)

$$\alpha_\theta^c = \mathbf{n} \sqrt{\frac{2}{3}} [g M^c - m] . \tag{3.26d}$$

The three surfaces are assumed to have similar shapes. Hence they obey similar equations. M_c, n^d, e_{c0}, λ and ξ are further material parameters.

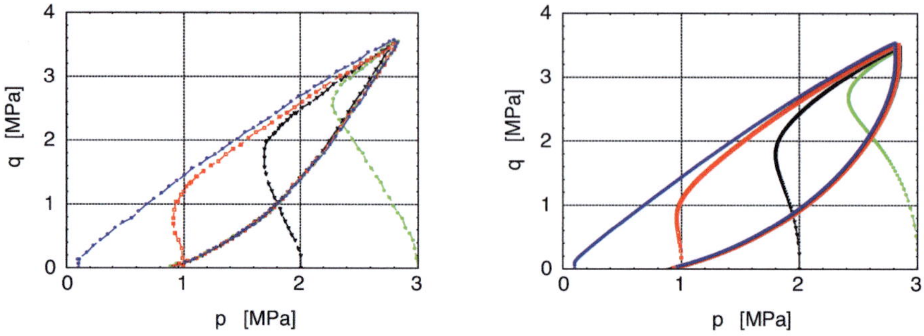

(a) Measurements (left figure) and simulations with SaniSand (right figure) of undrained monotonic triaxial tests

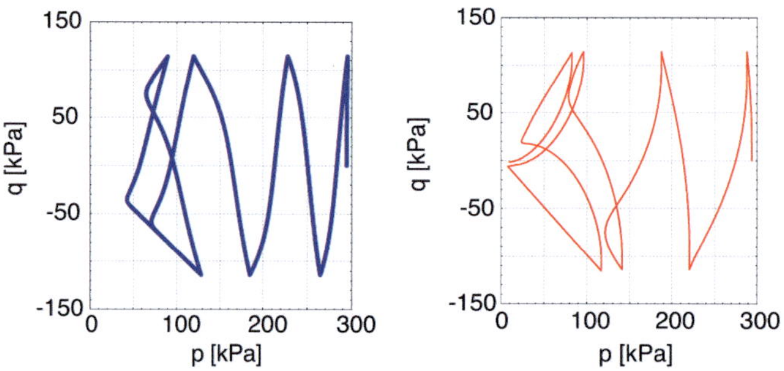

(b) Simulations of undrained cyclic triaxial tests with hypoplasticity and intergranular strain (left figure) and SaniSand (right figure)

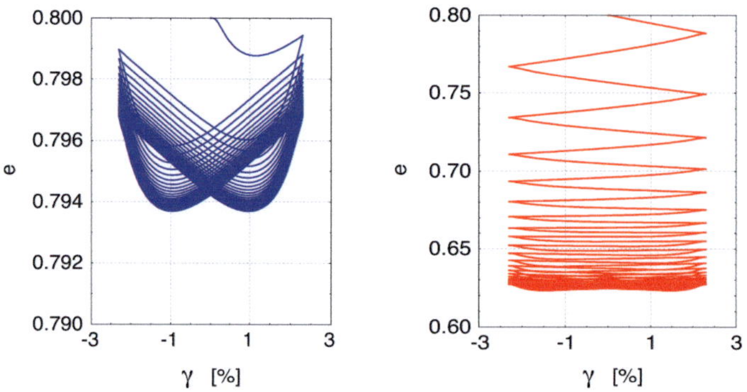

(c) Simulations of simple shear tests with hypoplasticity and intergranular strain (left figure) and SaniSand (right figure)

Figure 3.10.: Comparisons between measurements and simulations with SaniSand and hypoplasticity with intergranular strain (PRADA [122])

The model requires in total 15 material parameters.[5] Unfortunately, a detailed description of how to determine the parameters, has (to the authors knowledge) not been proposed by the authors Dafalias, Manzari and Taiebat.

The above presented equations have been implemented by Niemunis PRADA [122] in a user-subroutine for the FE-program ABAQUS [44]. He has performed several numerical simulations of laboratory element tests, in order to compare SaniSand with measurements as well as with the hypoplastic model with intergranular strain. Some results are shown in Figure 3.10.

Figure 3.10a presents measurements (left figure) and back-calculations with SaniSand (right figure) of undrained triaxial compression tests, performed by VERDUGO AND ISHIHARA [147]. The comparison reveals, that SaniSand can predict the observations well.

A comparison between SaniSand and hypoplasticity with intergranular strain is presented in Figure 3.10b. The shown simulations are back-calculations of undrained cyclic triaxial test, performed by ISHIHARA ET AL. [83]. Both models agree fairly well, but SaniSand reproduces a more realistic asymptotic butterfly.

Figure 3.10c shows another comparison between both models. The presented results correspond to a simple shear test with 50 load reversals. While the hypoplastic model with intergranular strain reaches a periodic response, SaniSand predicts an ongoing compaction without reaching an asymptote. This can be explained by the lack of a lower boundary, which was obtained in tests by Youd [154].

3.3. High cycle accumulation model

The models presented above are capable to predict cyclic loading behaviour of soils in laboratory tests and boundary value problems. However, these *implicit models*, wherein each cycle is traced numerically – i.e. the whole loop of a cycle is computed – can accumulate within big number of load reversals a significant numerical error due to the non-linearity of the model and the employed integration procedure. An alternative to implicit models is the so-called high cycle accumulation model proposed by NIEMUNIS ET AL. [113]. It is an *explicit model* which computes only average stress and strain accumulations. The model requires an underlying implicit model in order to compute strain amplitudes $\varepsilon^{\mathrm{ampl}}$. This can be every constitutive model suitable for the description of soil behaviour with load reversals.

In order to understand the high cycle accumulation model, some definitions are introduced first. A cycle, also referred to as loop, is understood as a recurring series of variables, such as loads, displacements, stresses or strains. A cycle i is described by its average value $\sqcup_i^{\mathrm{av}} =$

[5]Hypoplasticity with intergranular strain requires 13 material parameters.

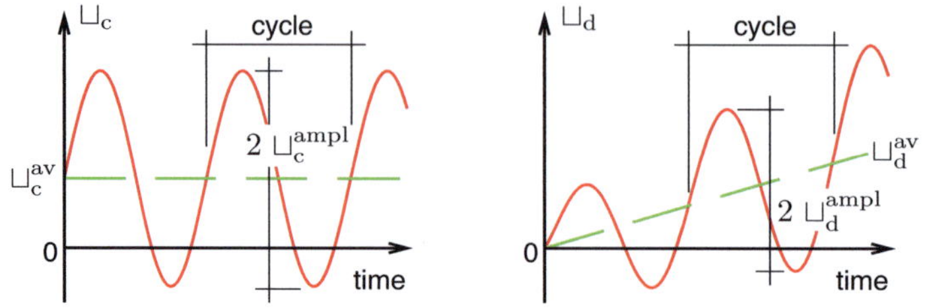

Figure 3.11.: Development of a variable under cyclic loading and its characteristic parameters

$(\sqcup_i^{max} + \sqcup_i^{min})/2$ and its amplitude $\sqcup_i^{ampl} = |\sqcup_i^{max} - \sqcup_i^{av}|/2$. Accumulation is the sum of residual differences between beginning and end of a cycle. The rate of a variable is its derivative with respect to the number of cycles $\dot{\sqcup} := \mathrm{d}\sqcup_d^{av}/\mathrm{d}N$ which is equivalent to $\dot{\sqcup} := \mathrm{d}\sqcup_d^{av}/\mathrm{d}t$ for a constant loading frequency. All terms are shown in Figure 3.11; on the left side a controlled variable $\sqcup_c$, on the right side a dependent (measured) variable $\sqcup_d$.

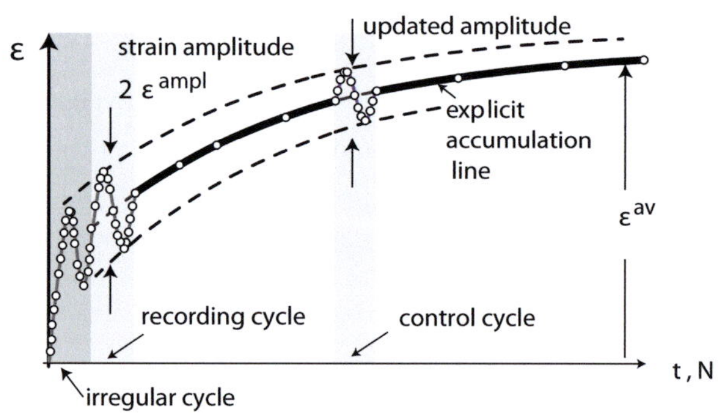

Figure 3.12.: Calculation procedure of the high cycle accumulation model (NIEMUNIS ET AL. [113])

Figure 3.12 (from NIEMUNIS ET AL. [113]) shows the interaction between the high cycle accumulation model and the underlying implicit model[6]. The first and second cycle is computed implicitly, but only the second one is considered in order to determine the cyclic strain amplitude. The first cycle is assumed to be irregular due a strong influence of the loading history. The subsequent cycles are taken into account explicitly with the high cycle accumulation model. If required, further implicit cycles can be computed with the underlying implicit model to check ε^{ampl} in order to correct the predicted accumulation rate if necessary. WICHTMANN [150] could show, that an update of the strain rate is for most cases not necessary for regular – i.e. simple, harmonic – cycles. This could be confirmed with own simulations.

The high cycle accumulation model has the basic form

[6] which, in frame of the presented work is the hypoplastic model with intergranular strain

$$\overset{\circ}{\mathbf{T}} = \mathsf{E} : (\mathbf{D} - \mathbf{D}^{\mathrm{acc}} - \mathbf{D}^{\mathrm{pl}}) \tag{3.27}$$

with the Jaumann stress rate tensor $\overset{\circ}{\mathbf{T}}$ being a function of an elastic stress-dependent stiffness E, the stretching tensor $\mathbf{D}$, the cumulative stretching $\mathbf{D}^{\mathrm{acc}}$ and an additional plastic stretching $\mathbf{D}^{\mathrm{pl}}$. The main part of the high cycle accumulation model is $\mathbf{D}^{\mathrm{acc}}$,

$$\mathbf{D}^{\mathrm{acc}} = \mathbf{m}\bar{f} = \mathbf{m}\, f_{\mathrm{ampl}}\, \dot{f}_N\, f_p\, f_Y\, f_e\, f_\pi \tag{3.28}$$

with an intensity factor $\bar{f}$ and a *cyclic flow rule* $\mathbf{m}$

$$\mathbf{m} = \frac{-\frac{1}{3}\left(p - \frac{q^2}{M^2 p}\right)\mathbf{1} + \frac{3}{M^2}\mathbf{T}^*}{\left\| \frac{1}{3}\left(p - \frac{q^2}{M^2 p}\right)\mathbf{1} + \frac{3}{M^2}\mathbf{T}^* \right\|} \tag{3.29}$$

which can be approximated either by the flow rule of the modified Cam-Clay model, Equation (3.29), or the hypoplastic flow rule Equation (3.9). The intensity factor $\bar{f}$ is the product of several influence factors:

- strain amplitude

$$f_{\mathrm{ampl}} = \begin{cases} \left(\frac{\varepsilon^{\mathrm{ampl}}}{\varepsilon^{\mathrm{ampl}}_{\mathrm{ref}}}\right) & \text{for} \quad \varepsilon^{\mathrm{ampl}} \leq 10^{-3} \\ 100 & \text{otherwise} \end{cases} \tag{3.30a}$$

- historiotropy, i.e. the dependence on previous cycles

$$\dot{f}_N = C_{N1}\left[C_{N2}\exp\left(-\frac{g^A}{C_{N1} f_{\mathrm{ampl}}}\right) + C_{N3}\right] \tag{3.30b}$$

  with

$$g^A = \int f_{\mathrm{ampl}}\, C_{N1}\, C_{N2} \exp\left(-\frac{g^A}{C_{N1} f_{\mathrm{ampl}}}\right) \mathrm{d}N$$

  which considers the number N of cyclic preloading and the amplitude.

- average mean pressure p^{av}

$$f_p = \exp\left[-C_p\left(\frac{p^{\mathrm{av}}}{p_{\mathrm{atm}}} - 1\right)\right] \tag{3.30c}$$

- average stress ratio η^{av} or $\bar{Y}^{\mathrm{av}}$

$$f_Y = \exp\left[C_y \bar{Y}^{\mathrm{av}}\right] \tag{3.30d}$$

$$\text{with} \quad \bar{Y} = \frac{Y - 9}{Y_c - 9} \quad Y = -\frac{I_1 I_2}{I_3} \quad Y_c = \frac{9 - \sin^2\varphi_c}{1 - \sin^2\varphi_c}$$

- void ratio e

$$f_e = \frac{(C_e - e)^2}{1+e} \frac{1+e_{\text{ref}}}{(C_e - e_{\text{ref}})^2} \tag{3.30e}$$

- changes of polarisation, which considers e.g. alternating large and small strain amplitudes or regularly varying changes of the loading direction.

$$f_\pi = 1 + C_{\pi 1}(1 - \cos\alpha) \tag{3.30f}$$

$$\text{with} \quad \dot{\alpha} = -C_{\pi 2}\,\alpha\,(\varepsilon^{\text{ampl}})^2$$

The plastic stretching $\mathbf{D}^{\text{pl}}$ is relevant in simulations of boundary value problems only, in which elements suffer stretching ($\mathbf{D} \neq 0$) due to deformations of neighbour elements, but no accumulated strains due to stress cycles ($\mathbf{D}^{\text{acc}} = 0$). For example, an element at ground level near a vertical cyclically loaded shallow foundation can have $\mathbf{D}^{\text{acc}} \approx 0$, but a considerable stretching $\mathbf{D}^{\text{pl}} \neq 0$, since the elements under the foundation contract. Equation (3.27) without the plastic stretching $\mathbf{D}^{\text{pl}}$ can generate decreasing stresses[7] which can even trespass the yield surface. The stretching $\mathbf{D}^{\text{pl}}$ projects the stresses back onto the yield surface and hence prevents inadmissible states; WICHTMANN [150].

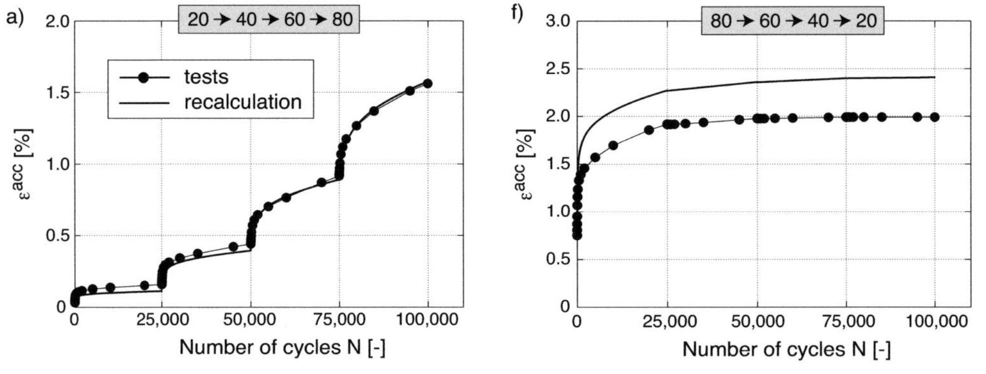

Figure 3.13.: Back-calculations of the cyclic triaxial tests with varying stress amplitude (WICHTMANN [150])

The model requires eleven additional material parameters which can be determined from cyclic triaxial tests. A detailed description of the parameter determination is given by NIEMUNIS ET AL. [113] and WICHTMAN [150]. Fortunately, some parameters can be assumed to be 1, respectively 0, if harmonic cycles are applied on a virginally loaded model or element. In

[7] pressure assumed positive

this case, neither a prehistory nor a polarisation due to complex cyclic loading paths has to be considered.

Figure 3.13 shows two back-calculations with the high cycle accumulation model of drained cyclic triaxial tests; WICHTMANN [150]. The samples were subjected to four packages of 25000 stress cycles, each with a constant amplitude. The test with continuously increasing amplitude for each package can be reproduced by the model almost perfectly. The simulation with continuously decreasing stress amplitudes, however, over-estimates the accumulated axial strain. Wichtmann explains the deviation by the strong influence of irregular cycles. If these cycles are omitted in the evaluation of the measurements and simulations, a better agreement between both is obtained (Wichtmann, personal communication).

4. Model tests with a cyclically loaded shallow foundation

Small-scale model tests were performed by H. Wienbroer and the author in order to understand the behaviour of cyclically loaded shallow foundations. The aim was to study the stabilisation behaviour during normal operation of the wind turbine subsequent to an intense load event. The measurements presented in this chapter serve mainly for a validation of the employed numerical models.

4.1. Model foundation

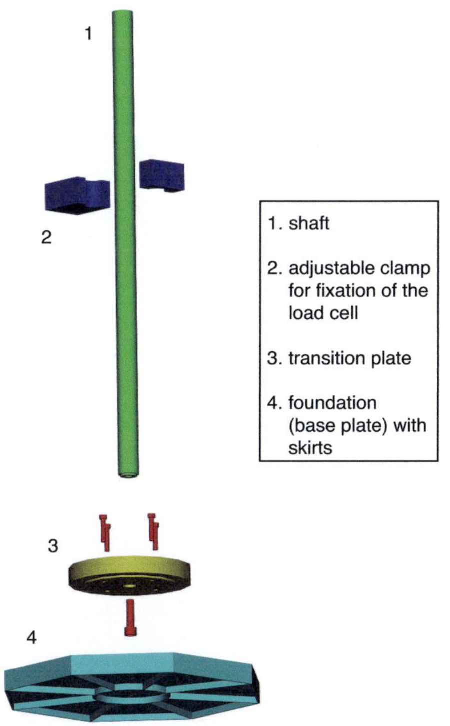

Figure 4.1.: Exploded view of the shallow foundation

An octagonal shallow foundation with short skirts was used in the model tests, as shown in Figure 4.1 and Appendix B. It has an outer diameter of $d = 28$ cm. The foundation is equipped with 0.66 cm long skirts, which divide the base plate into nine segments. On the topside of the base plate a transition plate is mounted. In the middle of that transition plate a round solid rod is attached by which the load is applied. The transition plate serves also as an adapter, which can is used to vary the dead weight of the structure in order to analyse different foundation dimensions. A vertical adjustable clamp with a load cell (type Z6 from HBM [66]) is attached at the shaft. The load cell serves to check the actually applied load, but it is not used to control the loading. Since the height of the clamp cannot be varied during a test, the ratio of horizontal load and moment at the base of the foundation remains constant. Figure 4.1 shows an exploded view of the foundation without load cell.

The base plate and shaft are made of aluminium, the transition plate consists of stainless steel (V2A). Due to the geometry and the small loads, the foundation may be considered as rigid. The inevitable bending of the base plate during the tests could not be observed. The total weight of the foundation including shaft, clamp and load cell amounts 5.1 kg.

A shallow foundation, consisting of one single plate with octagonal shape, has been chosen in accordance to successfully installed substructures for Offshore wind turbines (OWT); e.g. in the offshore wind parks NYSTED [117] and HORNSREV [72], both in Denmark. The foundation is simple to built and install and has been proven suitable in shallow waters with less than 10 m water depth. In areas with higher water depths, structures founded on a single plate seem to be uneconomical. Then, other geometries are expected to be more efficient. One solution will be discussed in Chapter 7. Details of a rough design of a shallow foundation are presented in Appendix A.

4.2. Test rig and instrumentation

Figure 4.2 shows a plot of the testing equipment and parts of the instrumentation. A former large triaxial cell, which was also used as a calibration chamber for penetration tests (e.g. CUDMANI [41] and CUDMANI AND STURM [42]), serves as basis. It is lined with a rubber membrane, which allows also model tests on saturated sand and, if required an additional back-pressure in order to guarantee complete saturation[1]. A measuring and a loading frame are mounted on top of the cell, on which displacement transducers and pneumatic cylinders, respectively, are attached. While the geometry of the frames provides general loading in two directions, Figure 4.2 shows, for sake of clarity, the instrumentation for test in one plane only. Detailed views of the test rig including photographs are shown in Appendix B.

[1] similar to triaxial tests and other laboratory tests with water saturated soil samples

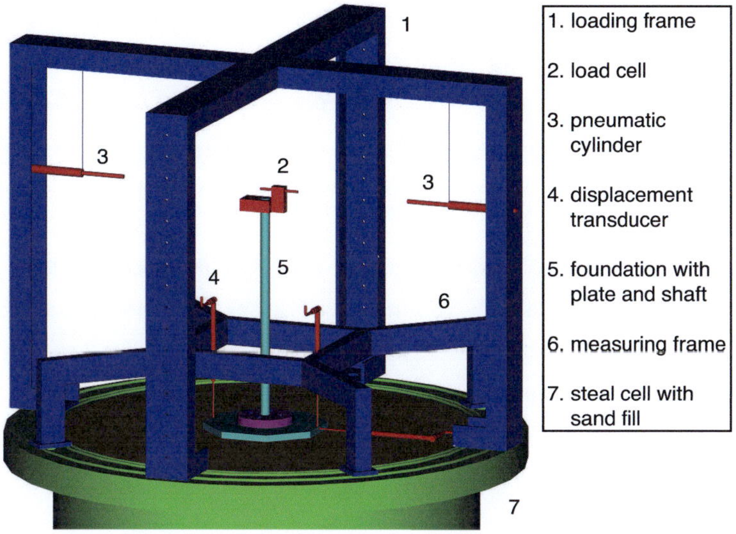

Figure 4.2.: Test arrangement with a shallow foundation on dense sand

4.2.1. Instrumentation

The load is applied by means of pneumatic cylinders (Airpel® M9). A pneumatic system was chosen in order to apply stress controlled cycles with small amplitudes. The cylinders are mounted at the loading frame with a hinge on one side and a soft tension spring on the other side which should prevent jamming and uncontrolled transverse forces. The pistons of the tensile loaded cylinders are connected with the load cell of the model foundation by means of thin and lightweight wires.

In total four cylinders can be mounted in the loading frame (item 3 in Figure 4.2). They are adjusted pairwise and rectangular to each other. This arrangement allows arbitrary symmetric and asymmetric load reversal in the plane spanned by the cylinders and load cell. It allows also to apply of an additional threshold pressure. All cylinders are subjected with the same initial pressure at rest. This has no effect on the foundation, but prevents jerky movements of the structure due to alternating release or rapidly taut wires during the tests. Hence, steady and smooth load reversals are possible.

The displacements of the foundation are measured with three, respectively, six WA-L HBM [66] displacement transducers. At least six are required for arbitrary loading, while three are sufficient for inplane loading, if translations and rotations perpendicular to the loading direction are neglected. The simplified arrangement with three transducers allows a direct evaluation of the displacements of the foundation from the measurements. Whereas a three-dimensional measurements of the displacements with 6 transducers, requires in any case a numerical error computation. The error is small if the degree of rotation of the transducers is small, which is the case if the distances between measuring and fixing points of a transducers are sufficiently apart. An error computation was not necessary for the presented tests, since the measured horizontal displacements are small compared to the vertical ones. I.e. the rotations of the transducers are negligibly small.

Earth pressure cells in the soil-structure contact zone have not been installed because of the small pressure of $\sigma_v' \approx 0.6$ kPa at rest. No suitable cells are available which can measure reliably such low stresses. It is intended in further tests to employ time-domain reflectometers (TDR) in order to measure indirectly the pressure in the contact zone by means of changes of the dielectric conductivity; e.g. SCHEUERMANN AND HUEBNER [130].

4.2.2. Model sand and preparation method

The cell was filled with a dry fine quartz sand. It has a uniformity coefficient of $U < 2$ and consists of rounded and sub-angular grains with a mean diameter $d_{50} = 0.14$ mm. The sand was placed in layers and compacted by means of a vibrator. The surface was levelled with a specially prepared vacuum cleaner. With a relative density of $D_r = 95\%$ of the prepared soil in the cell, the properties correspond those at typical offshore sites (compare Section 2.1).

4.3. Loading program and evaluation procedure

The foundation is placed on the sand bed carefully by hand. Thereby, the skirts penetrate into the soil, driven by the dead weight of the structure only. Prior to the actual loading test, the model foundation is subjected to symmetric pre-cycles with an amplitude of $F_{\text{pre}} = 0.05 \cdot F_{\text{ext}}$ in order to achieve full contact of the base plate with the sand bed. The pre-cycles re-densify the disturbed soil, owing to the installation, especially along the skirts.

In the presented tests a loading program consisting of a series of three equal load sequences, as shown in Figure 2.6 was applied on the foundation. It should represent three succeeding storms. The maximum load F_{ext} in all three sequences amounted 12 N and was applied in a height of 45 cm above ground level. F_{ext} corresponds the characteristic load assumed in the design and hence, is almost twice the design load used to determine the dimensions of the model foundation. This load has been chosen based on preceding model tests in which the load was carefully stepwise increased. The main criterion was the avoidance of an open gap. The maximum measured heaving of the foundation on the uplift side was 0.12 mm, and thus significantly smaller than the penetrated skirt lengths of 66 mm.

The ratio between the extreme load and the cyclic stress amplitude amounted $F_{\text{ext}}/F_{\text{cyc}}^{\text{ampl}} = 17\%$; i.e. $F_{\text{cyc}}^{\text{ampl}} = 2.04$ N. An average load was not applied; i.e. $F_{\text{cyc}}^{\text{av}} = 0$). This load combination and the magnitudes have been chosen based on a load table (Appendix C) determined for the offshore wind park *Borkum West*, which was kindly provided by the construction company *Ed. Züblin AG*. The cyclic load amplitude corresponds to a load during normal operation of a 5 MW OWT at a rather windy day.

The number of applied load cycles within a sequence was limited to $N_{\text{cyc}} = 500$. Preceding numerical predictions revealed, that this number is sufficient in order to observe a considerable amount of back-rotation of the model-foundation. The duration of a cycle was 5 seconds; i.e. a frequency of $f = 0.2$ Hz.

Between the three load sequences, as well as between the second and third load step[2] within a sequence, short resting periods of approximately 5 minutes were made. The extreme load F_{ext} was applied rapidly and kept constant for 5 minutes before it was removed again. This should show, if the soil under the foundation creeps under constant loading.

Figure 4.3 shows a sketch of the model foundation. The direction in which the extreme and cyclic loads were applied are drawn in. In the model-tests, presented in this work, F_{ext} and F_{cyc} act into the same direction, i.e. the angle in cross-section B-B amounts $\alpha = 0°$. A variation of the angle has been done only in the numerical study presented in Chapter 5.

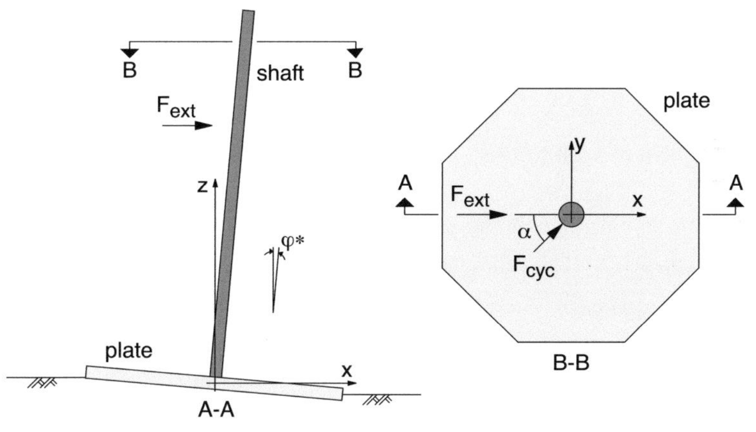

Figure 4.3.: Cross sections of the analysed foundation with loading point and load directions

The tilting of the foundation was measured – and calculated – in the direction of the maximum load, applied in the first step of the test sequence and is denoted with the angle φ, as shown in Figure 4.3. In order to compare simulations with model tests, the results were normalised, viz.

[2]a sequence consists of three load steps; compare Figure 2.6

$$\bar{\varphi}^{\text{av}} = 100 \cdot \left(\frac{\varphi_t^{\text{av}}}{\varphi_0^{\text{av}}} \right) [\%] \tag{4.1}$$

with φ_t^{av} denoting the current average rotation and φ_0^{av} denoting the rotation at the end of the first step of the first test sequence, i.e. for $N = 0$. This normalisation is actually not necessary, since the rotation is already a dimensionless variable and could be compared with other the simulations which have been done in prototype scale. However, Equation (4.1) is an appropriated representation to quantify the amount of back-rotation with respect to the residual rotation due to F_{ext}.

The settlement is normalised, viz.

$$\bar{s}^{\text{av}} = 100 \cdot \left(\frac{s_t^{\text{av}} - s_0^{\text{av}}}{\Delta s_{500}^{\text{av}}} \right) [\%] \tag{4.2}$$

with s_t^{av} denoting the current settlement and s_0^{av} denoting the settlement at the end of the first load step of a test sequence. The accumulated settlement after 500 load reversals in the third load step of the first test sequence is denoted as s_{500}^{av}. It has been introduced in Equation (4.2) in order to compare the amount of settlement of each test sequence. This representation allows a direct comparison of the relative settlement due to the cyclic load steps.

An alternative, and widespread, normalisation procedure, is to divide the current settlement with the width, respectively, diameter of the foundation. This ratio is an objective variable since it is independent of the dimensions of the foundation, as well as of the employed scaling factor in simulation and model test.

4.4. Results of the model tests

Measured rotations and settlement are presented in the following. A discussion of the results will be made in Chapter 6. The following figures show raw-data of the measurements; i.e. no noise reduction was applied.

4.4.1. Measured rotations

Figure 4.4 shows the imposed horizontal cyclic loads and Figure 4.5 the measured rotational response of the model foundation. The residual tilt at the end of the first extreme load was set to $\bar{\varphi}^{\text{av}} = 100\%$, according to Equation (4.1). One can see, that

- the foundation rotates back after an extreme loading; i.e. the structure straightens up again during cyclic loading with smaller amplitudes. It rotates in opposite direction to the previous tilt caused by the extreme loading;

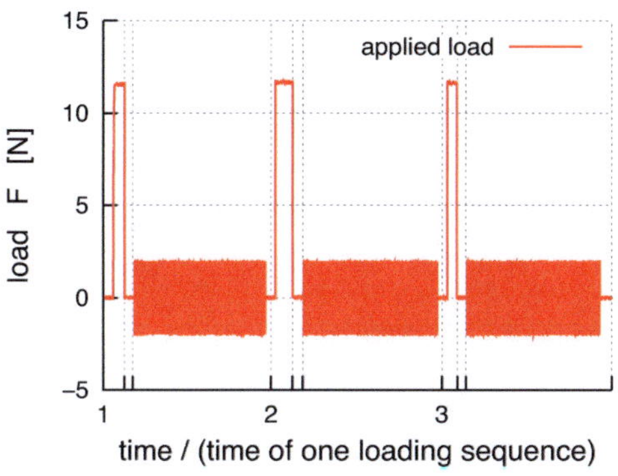

Figure 4.4.: Applied horizontal loads

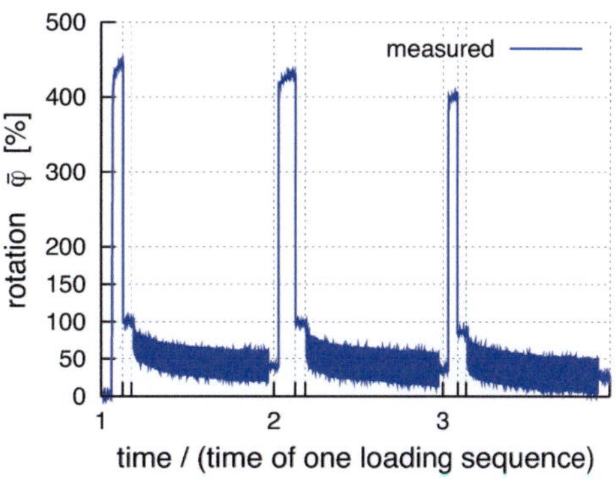

Figure 4.5.: Measured normalised rotation

- the amount of the *back-rotation* increases continuously with increasing number of applied load reversals. This is observed also after repeated extreme loading; i.e. the back-rotational behaviour can be reproduced;
- the maximum rotation $\bar{\varphi}_{\max}$ during extreme loading decreases for every repetition. The soil under the foundation develops obviously an increasing resistance against extreme loads of equal magnitude;
- the residual rotation $\bar{\varphi}_0^{\mathrm{av}}$ after an extreme loading decreases for every repetition;
- the average accumulated tilt $\bar{\varphi}_t^{\mathrm{av}}$ during cyclic loading in all three test sequences tends asymptotically to the same curve. I.e. the temporarily increased rotation due to the second and third extreme loading is swept out within a few cycles.

The decrease of maximum $\bar{\varphi}_{\max}$ and residual rotation $\bar{\varphi}_0^{\mathrm{av}}$ may be considered as a kind of *hardening*. The incremental stiffness of the soil seems to increase with an increasing number of load cycles and test sequences of equal load ratios.

A detailed view of the development of the tilt $\bar{\varphi}^{\mathrm{av}}$ during the three cyclic load packages of the test is shown in the Figure 4.7 to Figure 4.9. One can see, that the average tilt $\bar{\varphi}^{\mathrm{av}}$ decreases about linearly with the logarithm of time; i.e. its rate decreases with increasing number of load cycles. The back-rotation after 500 load reversals amounts in all three cyclic load steps approximately $\Delta\bar{\varphi}_{500}^{\mathrm{av}} \approx 60\%$. The amplitude of rotation increases slightly in the cyclic load step of the first load sequence and remains constant in the subsequent cyclic load steps.

An important issue is the creep behaviour of the fine grained material used in the model tests. In general, viscous effects of granular soils are neglected for practical purposes, although creep and relaxation of sand could be observed and measured in laboratory element tests; e.g.

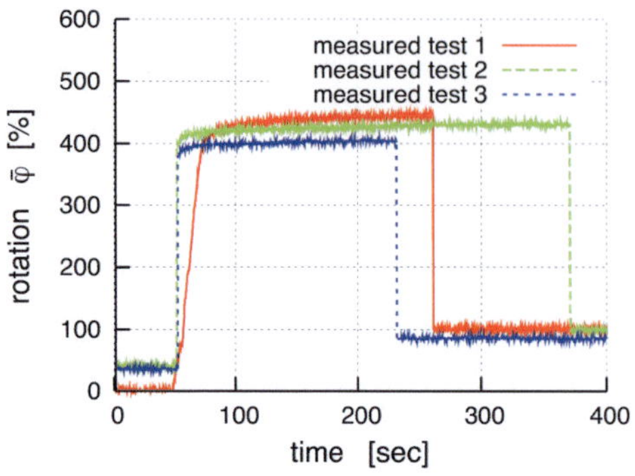

Figure 4.6.: Measured creep behaviour during the three extreme loading

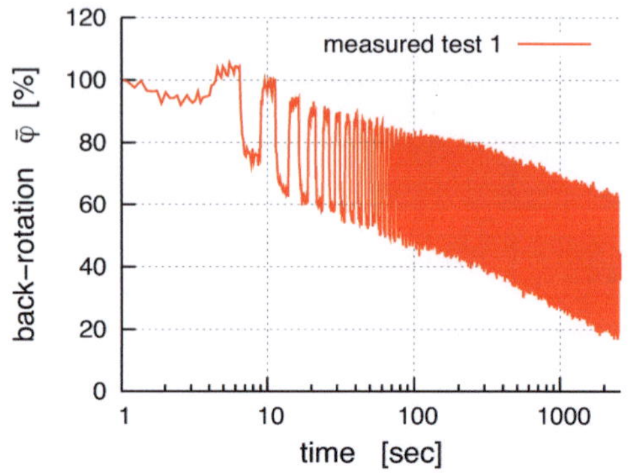

Figure 4.7.: Accumulated rotation in the first cyclic load package

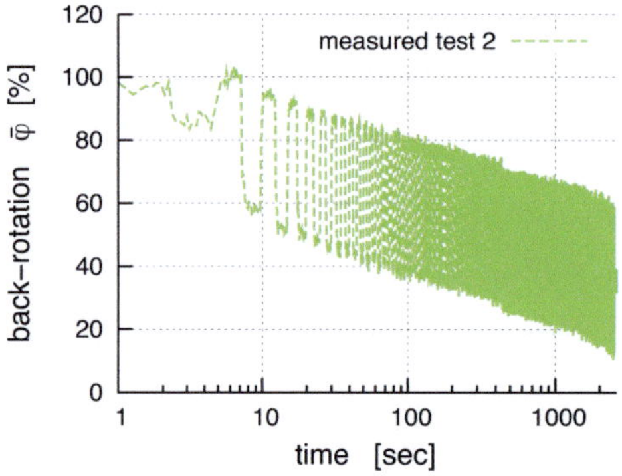

Figure 4.8.: Accumulated rotation in the second cyclic load package

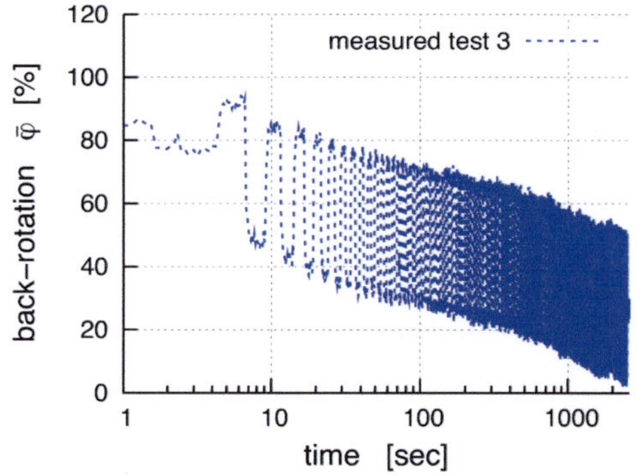

Figure 4.9.: Accumulated rotation in the third cyclic load package

MATSUSHITA ET AL. [99]. Figure 4.6 shows that also the model sand under the applied loading conditions can be considered this way. During constant static load the maximum rotation remains also constant and does not increase as it would if the soil would tend to creep.

4.4.2. Measured settlement

Figure 4.10 shows the measured average vertical displacement of the model foundation. The settlement during the first cyclic load step was set to again to 100%, according to Equation (4.2). A detailed view of the settlement on compression and uplift side is shown in Figure 4.11. One can see, that

- the foundation sinks continuously into the subsoil. The rate of settlement decreases with increasing number of load reversals;

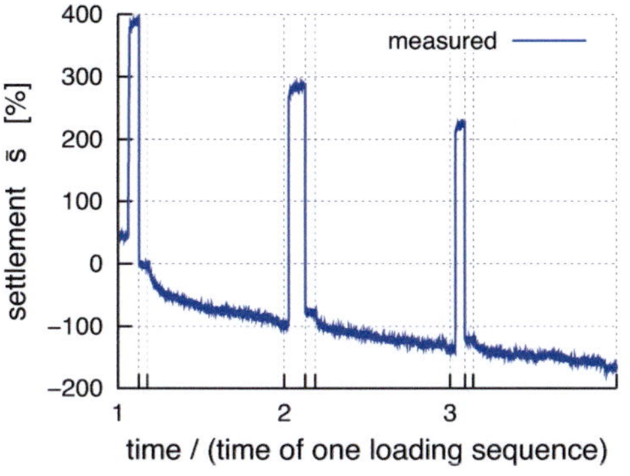

Figure 4.10.: Normalised measured vertical displacements of the model foundation

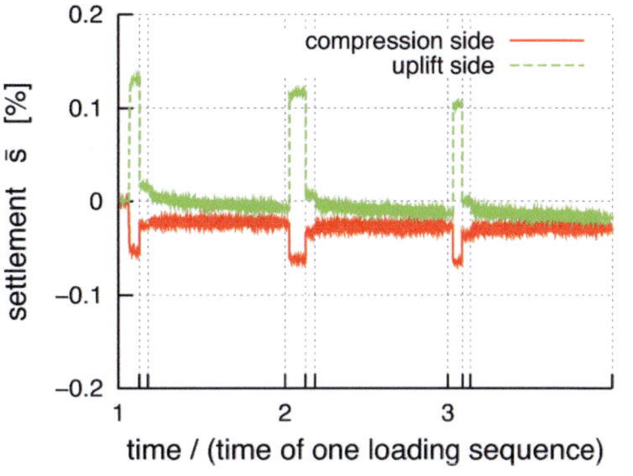

Figure 4.11.: Absolute vertical displacements of the model foundation on uplift and compression side

- the accumulated settlement during cyclic loading tends asymptotically to the same settlement curve for all three load sequences. This agrees with the behaviour observed for the accumulated rotation;
- similar to Figure 4.5 the loading history, i.e. the second and third extreme loading, is also wiped out within a few load reversals.

From Figure 4.11 it becomes apparent, that the settlement during cyclic loading on the uplift side of the foundation dominates the overall settlement behaviour, while the settlement on the compression side can be almost neglected. The response of the soil on the compression side during extreme loading seems to be quasi-elastic.

5. FE-simulations of a substructure founded on one large plate

In order to understand the mechanisms and to identify the influencing parameters of the rotational behaviour during cyclic loading, a back-calculation of the model tests and a numerical parameter study has been done. For this purpose, constitutive models, presented in Chapter 3, were applied, which have been proved suitable to describe the soil behaviour in simulations of boundary value problems realistically; e.g. KARCHER [87] and SŁOMIŃSKI [134]). The numerical simulations will be used in Chapter 6, in conjunction with the results of the model test (Chapter 4), for a discussion and identification of the governing mechanisms of stabilisation.

5.1. Model description

The numerical simulation of the model tests and the subsequent parameter study has been done by means of the *finite element method* (FEM). The commercial program ABAQUS [44] was used, since the employed constitutive models have been implemented in user-subroutines for this code and were kindly provided by A. NIEMUNIS[1].

5.1.1. FE-Models

Finite element meshes for shallow foundations on a half space are rather simple geometries. They can be generally generated within a short time. More time-consuming are 3d meshes and foundations with skirts and multiple interfaces, i.e. soil-structure contact surfaces. The preprocessing has been done by means of the commercial program *Hypermesh*, which is part of the software suite *Hyperworks* from ALTAIR ENGINEERING [6].

The main difficulty in numerical simulations of geotechnical problems is the correct modelling of the soil behaviour. The typical stress-strain relation of soils governs element type and shape, mesh geometry as well as the required number of elements. In particular, tension and localised distortions have to be obviated.

Two different finite element models were used in the parameter study; a simplified plane strain and a 3d model. The latter was used to validate the simplifications made in the plane strain model and to analyse the influence of the cyclic loading direction, shown in Figure 4.3.

[1] http://www.pg.gda.pl/~aniem/index.html

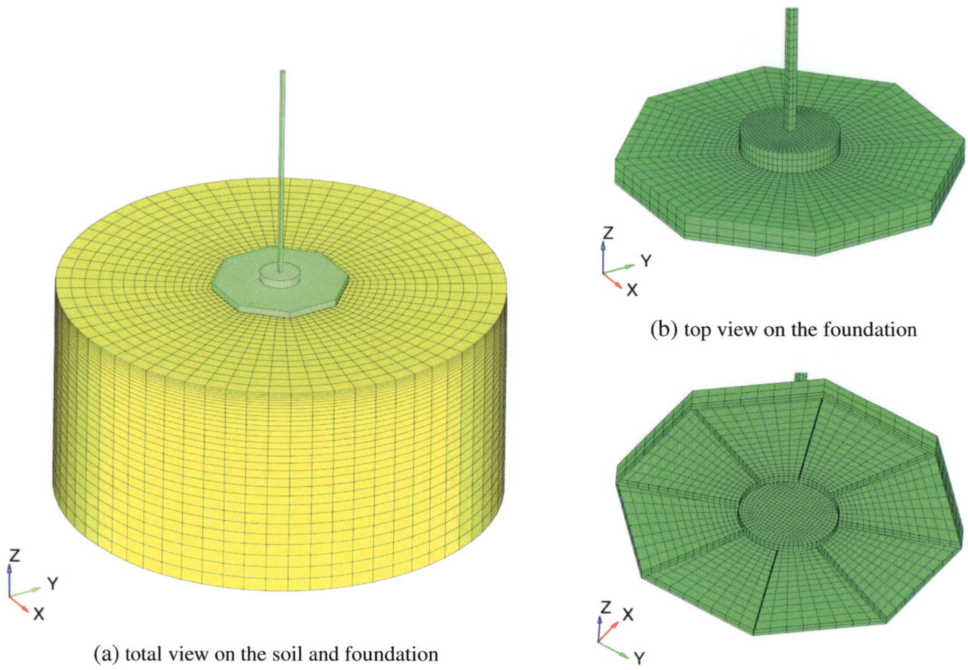

(a) total view on the soil and foundation

(b) top view on the foundation

(c) bottom view on the foundation

Figure 5.1.: 3d finite element mesh and partial views of the foundation

Both FE-models are scaled by 1:125 with respect to the dimensions of the model tests presented in Chapter 4. The diameter of the FE-model of 125 m is given by the dimension of the cell. The modelled height of the soil column amounts $\approx$ 60 m, which corresponds a third of the actual soil column height in the test, is two and a half times the foundations diameter. This is justifiable since the maximum loading is far below the ultimate bearing capacity. A conventional failure mechanism consisting of plasticized zone which may reach the boundaries, can be excluded. Hence, the boundary may not affect the results of the numerical simulations. A tangential frictionless sliding was allowed at the vertical and horizontal boundaries of the modelled soil column.

Figure 5.1 shows one of the employed 3d models. A double-symmetrical mesh with cubical 8-node elements[2] was chosen. The edge length of the soil elements close to the foundation is between 25 and 90 cm. The maximum *aspect ratio* is 3.7, but generally between 1.3 and 2. The elements have *jacobian values*[3] between 0.859 and 0.998. The elements have almost no *warpage*. This parameter describes the *warpage* of one side of an element in a 3d-space. It is

[2]in terms of ABAQUS: C3D8 and C3D16P

[3]the value ranges from 0.0 to 1.0. A perfectly shaped rectangular element, has a value of 1.0.

a measure of the vertical distance of a node to a plane spanned by the remaining 3 nodes. A critical parameter is the maximum angle of an element. Acute and obtuse angles have to be avoided in order to allow a stable numerical simulation. Due to the chosen elements and the given geometry, twelve elements per layer had a maximum angle of 135°. These elements are, however, sufficiently far away from the skirts, and hence experience almost no or only very small distortions.

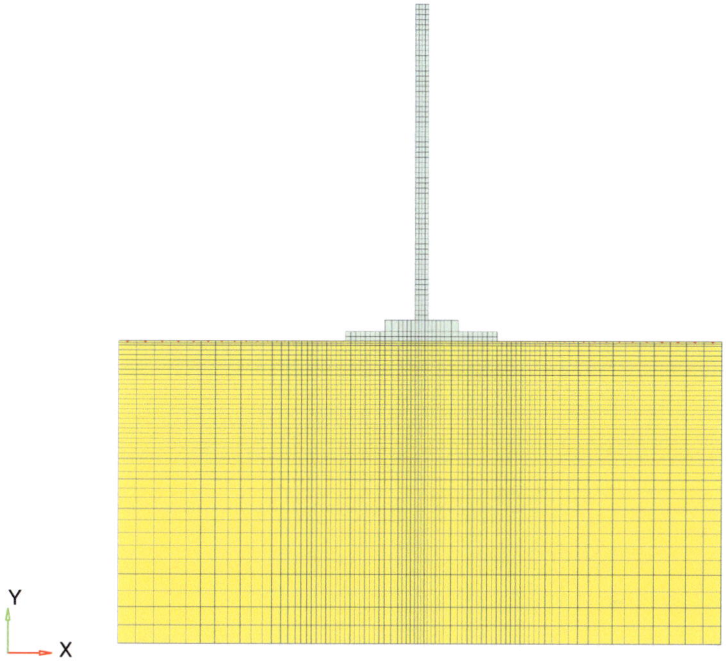

Figure 5.2.: 2d finite element mesh

A numerical convergence analysis was done in order to study the influence of element size, mesh geometry, element type and increment length for the numerical integration on the results. From these findings, an appropriate model was chosen and used in both, the simulation of the model tests and the parameter study. By this approach artificial numerical influences on the results can be avoided as much as possible. The chosen mesh is shown in Figure 5.2. Details of the convergence study are presented in Appendix D.

The quality of a plane 2d mesh is generally better than the one of a 3d model. The *aspect ratio* was always less than 2 with exception of a few elements on the bottom edge in the middle of the mesh. The 2d-elements of a plane-strain model have obviously no *warpage*. The angle of the elements amounted always 90°.

In order to account for the soil-structure interaction, several approaches were analysed. The interfaces were arranged at three different locations in the model:

1. the foundation rests on top of the sand bed, while skirts and bedding are not considered;

2. the soil between the skirts is replaced by an elastic material with the same weight of the replaced soil and is added to the foundation. The bedding depth is assumed to be equal to the length of the skirts;

3. skirts as well as base plate are coated by contact surfaces. The complete geometry of the foundation is modelled.

All three approaches were additionally computed without contact surfaces, i.e. the foundations elements were pinned at the soil elements. Hence, neither a relative displacement nor an open gap between both was allowed.

Since the obviation of an open gap was preconditioned in the design, the results of the numerical simulations were almost identical for all models. Neither the location of the interfaces nor the other mesh details had an significant influence on the results.

5.1.2. Constitutive parameters

The used models are described in more detail in Chapter 3. In most simulations, the *hypoplastic model with intergranular strain* (hypo+igs) in combination with the *high cycle accumulation model* (hca) was employed. Other constitutive models were used only in order to compare the simulations and to support the plausibility of the numerical results.

The determination of the hypoplastic parameters of the model sand has been done in the framework of a master thesis; UTUMI [145]. The intergranular strain parameters were roughly estimated on the basis of own experiences. The parameters of the high cycle accumulation model were determined based on an comparison of the model sand with other sands for which Wichtmann has already determined the parameters (not published yet). He distinguished the analysed materials by their grain-size-distribution curves. In order to validate the adopted parameters of the high cycles accumulation model, UTUMI [145] performed several supplementing laboratory tests. The parameters are listed in Table 5.1.

This rough estimation is sufficient, since only a qualitative back-calculation of the model tests was intended. A quantitative back-calculation is out of the scope of the constitutive model, due to the small stresses in the model tests. The vertical pressure of the model foundation at rest was only 0.6 kPa.

Table 5.1.: Constitutive parameters of the model sand for *hypo+igs* and *hca*

hypoplasticity with intergranular strain:

h_s	n	e_{c0}	e_{d0}	e_{i0}	φ_c	α	β	m_T	m_R	R_{max}	β_χ	χ
625 MPa	0.33	1.05	0.67	1.21	32.8°	0.18	1.12	3.5	6.0	$1 \cdot 10^{-4}$	0.2	6

high cycle accumulation model:

C_{N1}	C_{N2}	C_{N3}	C_P	C_{ampl}	C_Y	C_e	$C_{\pi 1}$	$C_{\pi 2}$	p_{atm}	e_{ref}	$\varepsilon_{\mathrm{ref}}^{\mathrm{ampl}}$
$1.21 \cdot 10^{-3}$	0.39	$5.0 \cdot 10^{-5}$	0.44	1.0	2.05	0.52	4.0	200	100 kPa	0.901	$1.0 \cdot 10^{-4}$

The comparison of the hypo+igs model and *Sanisand* (sani) in Section 5.4 was done with the parameters of Toyuora sand. The parameters for Sanisand were taken from DAFALIAS [43], the one for hypo+igs from CUDMANI [41]. Both sets are listed in Table 5.2.

Table 5.2.: Constitutive parameter of Toyoura sand for *hypo+igs* and *sani*

hypoplasticity with intergranular strain:

h_s	n	e_{c0}	e_{d0}	e_{i0}	φ_c	α	β	m_T	m_R	R_{max}	β_χ	χ
120 MPa	0.69	0.98	0.61	1.13	32°	0.12	1.0	2.0	5.0	$1 \cdot 10^{-4}$	0.1	2.0

Sanisand:

G_0	ν	M	c	λ_c	e_0	ξ	m	h_0	c_h	n^b	A_0	n^d	z_{max}	c_z
125	0.05	1.25	0.712	0.019	0.934	0.7	0.01	7.05	0.97	1.1	0.704	3.5	4	600

5.2. Back-calculation of model tests

In order to validate the numerical model, the model tests presented in Chapter 4 were back-calculated with a scaled 2d model in prototype dimensions. Since the actual applied load in the model tests was twice the design load $F_{\mathrm{ext,d}} \approx 0.5 \cdot F_{\mathrm{ext,c}}$, used in Appendix A for the determination of the dimensions of the foundation, the used FE-model incorporated contact surfaces between soil and foundation. However, an open gap could neither be observed in the model tests nor in the simulations.[4]

[4] An open gap could have been detected in the tests only indirectly, by means of the measured vertical displacement on the uplift and compression side of the foundation

The results of the back-calculation are presented in Figure 5.3. Figure 5.3a shows the normalised rotation according to Equation (4.1) for both, the design load and the actually applied load, while Figure 5.3b shows the settlement on the uplift and compression side for the actually applied load only. The cyclic load step of the first loading sequence was calculated with *hypo+igs* and *hca*, the second one with *hypo+igs* only. In order to reduce an accumulated numerical error due to the implicit calculation procedure of the hypoplastic model, in which every loop is traced, only the first 25 cycles were computed of the cyclic step of the second test sequence. The third sequence was not simulated. Like in the model tests before, the residual rotation after unloading from the first extreme load was set to $\bar{\varphi} = 100\%$.

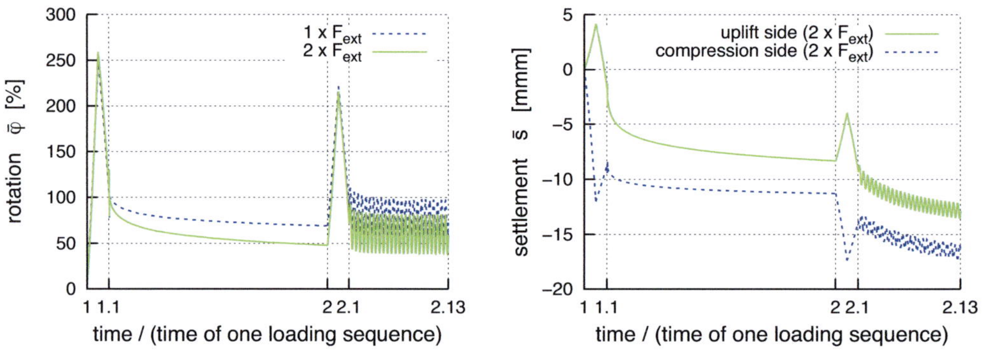

(a) Normalised rotation (b) Absolute settlement

Figure 5.3.: Back-calculation of the model test with *hca* and *hypo+igs* (first cyclic step) and *hypo+igs* (second cyclic step)

A comparison of Figure 5.3a with 4.5 reveals that all relevant observations made in the model tests can be reproduced by the numerical simulations:

- a back-rotation is obtained for both models, *hypo+igs* with and without *hca*;
- the maximum rotation $\bar{\varphi}_{\max}$ as well as the residual rotation $\bar{\varphi}_0$ after unloading from the extreme load, decrease for repeated extreme loads;
- the rate of back-rotation $\dot{\bar{\varphi}}^{\text{av}}$ decreases about linearly with the logarithm of the number of cycles.

The *hardening* behaviour, i.e. the increase of the incremental stiffness, observed in the model tests is also obtained in the simulations. However, both, the maximum rotation $\bar{\varphi}_{\max}$ as well as the relative back-rotation $\Delta\bar{\varphi}_{500}$[5] are predicted to be smaller than actually measured. A comparison of the ratios shows a better agreement of simulations and measurement, viz.

$$\left(\frac{\Delta\bar{\varphi}_{500}}{\bar{\varphi}_{\max}}\right)_{\text{meas}} \approx \left(\frac{\Delta\bar{\varphi}_{500}}{\bar{\varphi}_{\max}}\right)_{\text{sim}} \approx 0.17 \tag{5.1}$$

A comparison of predicted and measured settlement can hardly be done, even if the results are divided by the diameter of the foundation. In the simulations, the absolute values are larger and the dilatancy smaller, due to the higher effective vertical stresses but equal initial density. Despite of this difficulty, a comparison of Figure 5.3b with Figure 4.11 reveals, that simulation and measurements are qualitatively similar. A different behaviour can be seen for the predicted heave of the foundation on the uplift side during extreme loading. Compared to the test it is significantly smaller, hence the average settlement, which is the sum of both curves shown in Figure 5.3b, is overestimated in the simulations. Besides of the scaling factor, several further explanations for this discrepancy are possible.

It has been shown in Section 3.1.1 that *hypo+igs* underestimates the dilatancy of very dense sand samples under shearing; STURM [137]. Densities close to e_d can hardly captured by the model. *Hypo+igs* is more suitable for loose and medium dense sand with a maximum relative density of $D_r = 90\%$.

Another reason for the discrepancy can be found in the employed FE-model. Shear-induced dilatancy depends on the discretisation, i.e. on the element size; REBSTOCK [125, 126]. Shear deformations are concentrated in one element, or element row, and cannot be transfered into adjoining elements, since this would require an additional rotational degree of freedom of the finite-elements. Such a degree of freedom is incorporated in polar models, such as the *Cosserat* model, e.g. GRAMMENOUDIS [60].

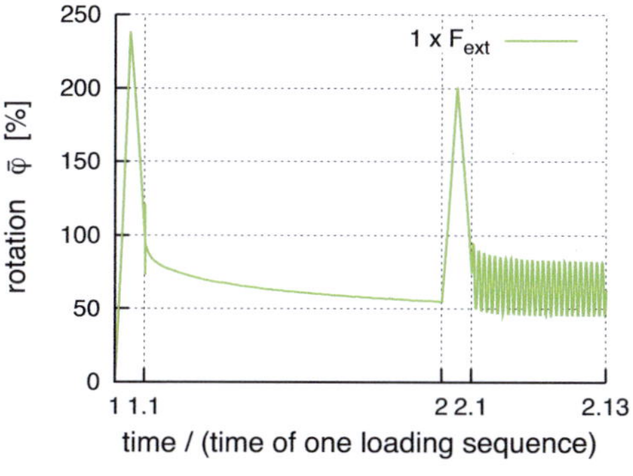

Figure 5.4.: Back-calcualtion of model test without contact surfaces between foundation and soil

[5] the accumulated back rotation in the first 500 cycles

Since some simulations, including the predictions, were actually done prior to the model tests, FE-models which do not incorporate contact surfaces were also employed. Figure 5.4 shows the result with one of these models. The extreme load, applied in this simulation corresponds to the design load $F_{\text{ext}} = N^3 \cdot 6\ \text{N} \approx 9681.85$ kN. That is half of the actually applied load. The rotational response of the foundation agrees also quite well with the measurements. Compared to Figure 5.3a, which presents the results of the simulation with the same load but incorporating interfaces, $\bar{\varphi}^{\text{av}}$ is somewhat smaller and $\bar{\varphi}^{\text{av}}_{500}$ slightly larger than without interfaces.

Summarising, one can say that the FE-model is capable to describe qualitatively the stabilisation behaviour, i.e. the back-rotation and the re-compaction[6]. The absolute values are not quite obtained, since neither the constitutive parameters have been determined reliable, nor does the FE-model corresponds to the dimensions of model test. A simulation in model dimensions is out of the scope of *hypo+igs* anyway. But, the results of the back-calculation reveal, that the FE-model is suitable for a qualitative numerical study on the influence of parameters on the stabilisation behaviour.

5.3. Parameter study

In order to analyse the influence of certain factors on the back-rotation, a variation of the amplitude and average value of the cyclic loading as well as of the initial relative density was numerically executed. The values of the varied parameters are listed in Table 5.3. The simulations were done with the FE-model, which was used for the back-calculation of the model tests. The extreme load was applied once and had always the same value and direction. Although up to 100000 cyclic load reversals were computed, only the first 10000 cycles were evaluated, since all relevant details can be derived from these results. If a foundation is capable to back-rotate from a tilted state due to a previously occurred extreme loading, the major part of the stabilisation process should be accomplished. Otherwise, the back-rotation is insufficient and should not be considered in the design.

Table 5.3.: Values for the numerical parameter study. The loads are referred to the extreme load F_{ext}

Quantity	**Variations**
F^{av}	-15%, -7.5%, $\pm 0\%$, $+7.5\%$, $+15\%$
F^{ampl}	5%, 10%, 15%, 20%
D_r^{ini}	65%, 75%, 85%, 95%

[6] this will be discussed in detail in Chapter 6

In order to compare the numerical results, the amount of accumulated back-rotation is normalised, viz.

$$\hat{\varphi}^{\mathrm{av}} = 100 \cdot \left(\frac{\varphi_0^{\mathrm{av}} - \varphi_t^{\mathrm{av}}}{\varphi_0^{\mathrm{av}}} \right) \cdot \left[\left(\frac{\Delta\varphi^{\mathrm{av}}}{\varphi_0^{\mathrm{av}}} \right)_{\mathrm{ref}} \right]^{-1} [\%] \tag{5.2}$$

with φ_0^{av} denoting the residual rotation at the end of the second load step and φ_t^{av} the current rotation of the foundation. The angle $\Delta\varphi^{\mathrm{av}}$ describes the sum of accumulated rotation in the first 10000 cycles, viz.

$$\Delta\varphi^{\mathrm{av}} = \varphi^{\mathrm{av}}_{(N=10000)} - \underbrace{\varphi^{\mathrm{av}}_{(N=0)}}_{\varphi_0^{\mathrm{av}}}$$

Equation (5.2) considers a reference model with $F^{\mathrm{av}}_{\mathrm{ref}} = 7.5\% \cdot F_{\mathrm{ext}}$, $F^{\mathrm{ampl}}_{\mathrm{ref}} = 10\% \cdot F_{\mathrm{ext}}$ and $D_r = 85\%$. This load combination was chosen as reference model, since this loading condition is most likely to occur after an extreme load event, according to the load table shown in Appendix C.

The settlement is likewise normalised with the same reference model, viz.

$$\hat{s}^{\mathrm{av}} = \left(\frac{s_t^{\mathrm{av}} - s_0^{\mathrm{av}}}{\Delta s^{\mathrm{av}}_{\mathrm{ref}}} \right) \tag{5.3}$$

In addition to amplitude, average value and initial relative density, the influence of the relative loading direction of the cyclic load was analysed. In Table 5.4 are listed the applied values of the parameter α, which denotes the relative direction of the cyclic load referred to the extreme load; see Figure 4.3. Since a 3d simulation had to be performed for this variation, only the load combination of the reference model was analysed.

Table 5.4.: Varied cyclic loading directions

Quantity	**Variations**
direction α	0°, 45°, 90°, 135°

A variation of the constitutive parameters was not done, since neither airtight model tests for a quantitative back-calculation are available, nor a precise parameter determination were done for the model sand. Sensitivity studies of the employed constitutive models were done by means of simulation of laboratory element tests for the hypoplastic model by HERLE [67] and for the high cycle accumulation model by WICHTMANN [150].

5.3.1. Rotational behaviour

We first look at the rotational behaviour, since this is the main object of interest. The settlement behaviour is discussed in the next subsection, but only as far as necessary for the subsequent discussion of the mechanisms of stabilisation.

5.3.1.1. Influence of the average value F^{av}

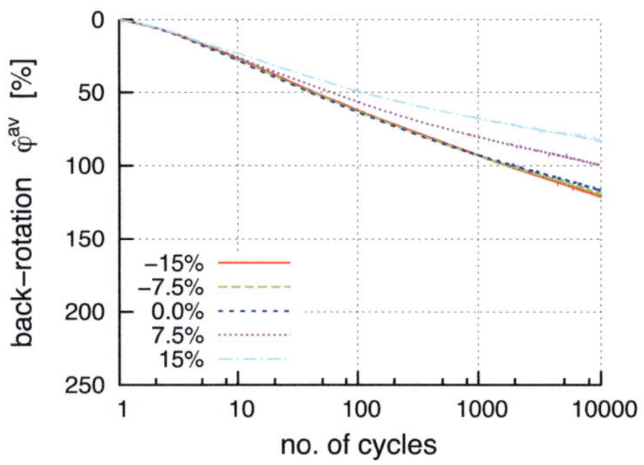

(a) $D_r = 85\%$, $F^{\mathrm{ampl}} = 10\%$

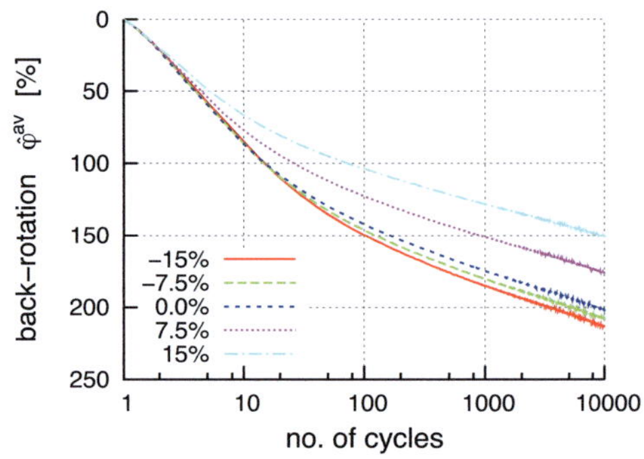

(b) $D_r = 85\%$, $F^{\mathrm{ampl}} = 20\%$

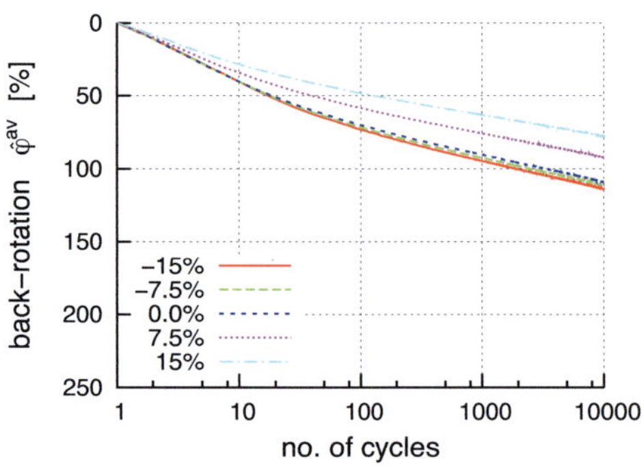

(c) $D_r = 65\%$, $F^{\mathrm{ampl}} = 10\%$

Figure 5.5.: Influence of F^{av} on $\hat{\varphi}^{\mathrm{av}}$

Figure 5.5 shows the normalised development of the rotation during the first 10000 cycles for different average values F^{av}, but equal initial relative density D_r^{ini} and amplitude F^{ampl}. A back-rotation is obtained for all F^{av}. The influence of the average value is small and decreases slightly with increasing F^{av}. This is reasonable because a negative value actually means that the foundation is pulled back, since $-F^{\mathrm{av}}$ acts in opposite direction to F_{ext}.

The development of the rate of back-rotation is somewhat different from the one in the model tests. In simulations with larger amplitudes and lower relative densities, a kink in the curve between the 10th and 30th cycle can be observed in the half-logarithm representation of the average rotation. This could also be seen in the model tests, but less distinctive. The inclination of the rotation curves is over-proportionally larger for the first cycles and smaller for higher number of cycles, but in any case linear apart from the kink. The distinctness of the different inclinations, as well as the location of the transition point, shifts to a higher number of cycles with decreasing average load. In Chapter 6 we will see, that the different inclinations indicate different mechanisms which govern the back-rotation.

5.3.1.2. Influence of the amplitude F^{ampl}

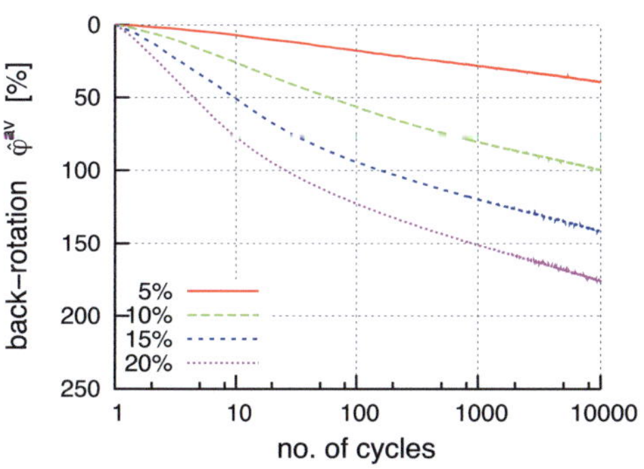

(a) $D_r = 85\%$, $F^{\text{av}} = +7\%$

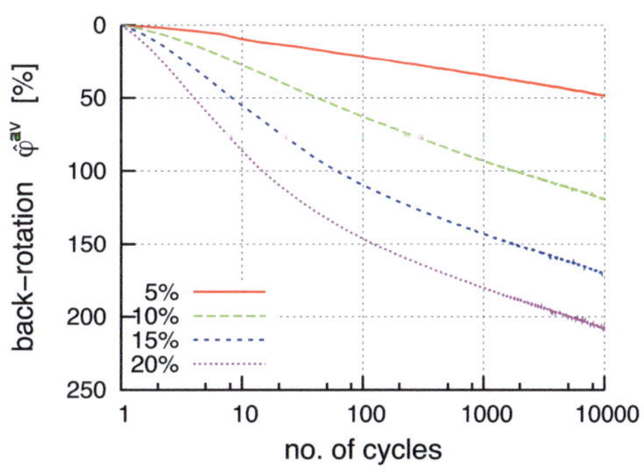

(b) $D_r = 85\%$, $F^{\text{av}} = -7\%$

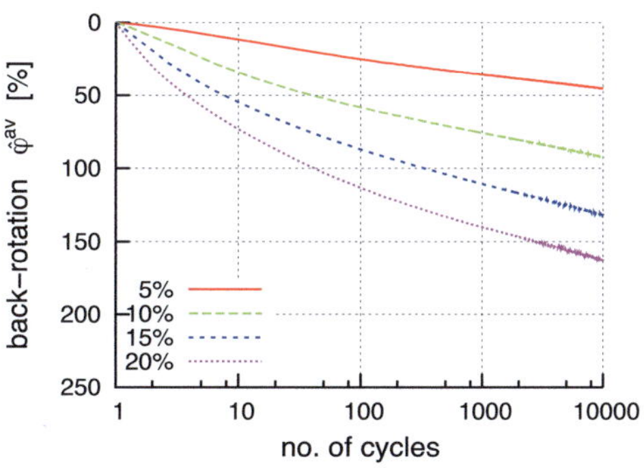

(c) $D_r = 65\%$, $F^{\text{av}} = +7\%$

Figure 5.6.: Influence of F^{ampl} on $\hat{\varphi}^{\text{av}}$

A comparison of simulations with different amplitudes, but constant initial density and average value, shown in Figure 5.6, reveal, that the accumulated back-rotation for a given number of cycles increases significantly with increasing amplitude. The increase with F^{ampl}, however, is under-proportional. This indicates a stabilisation and an asymptotic approximation of a maximum attainable back-rotation. That means, the back-rotation is most likely limited by a lower bound.

Like in Figure 5.5 before, a kink in the curve of the accumulated back-rotation can be observed. Distinctness and location of the kink increases and shifts to higher number of cycles, but vanishes for decreasing amplitudes, i.e. $F^{\mathrm{ampl}} = 5\%$.

5.3.1.3. Influence of the initial relative density D_r

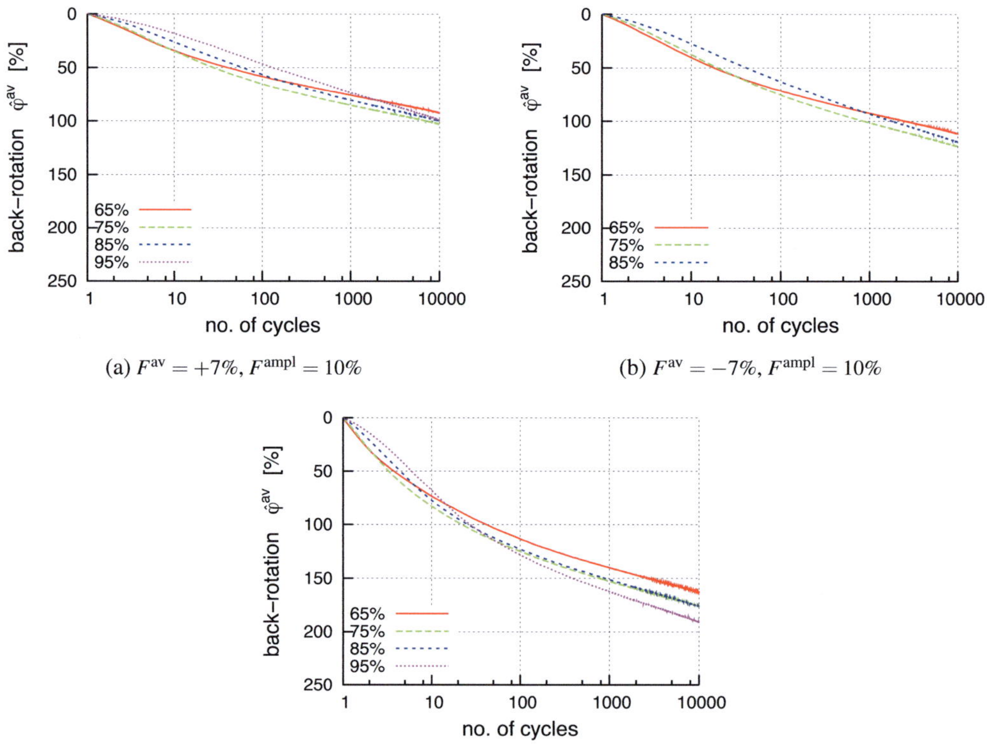

(a) $F^{\mathrm{av}} = +7\%$, $F^{\mathrm{ampl}} = 10\%$

(b) $F^{\mathrm{av}} = -7\%$, $F^{\mathrm{ampl}} = 10\%$

(c) $F^{\mathrm{av}} = +7\%$, $F^{\mathrm{ampl}} = 20\%$

Figure 5.7.: Influence of D_r on $\hat{\varphi}^{\mathrm{av}}$

Figure 5.7 shows the dependency of the back-rotation on the initial relative density D_r. One can see, that D_r can be neglected for small amplitudes, but becomes considerable for larger amplitudes. The accumulated back-rotation decreases with decreasing initial relative density.

Simulations of foundations on sand with $D_r^{\text{ini}} < 65\%$ show, that the back-rotation becomes even smaller and can get negligible for foundations on loose sand.

The normalised back-rotation according to Equation (5.2), shown in Figure 5.7, suggests that foundations on medium dense and dense sand are equipollent with respect to the amount of back-rotation. However, Equation (5.2) says only, that the relative back-rotation $\Delta\varphi^{\text{av}}$ referred to the residual rotation at the beginning of the cyclic load step φ_0^{av} is equal. But since φ_0^{av} increases with decreasing relative density D_r, the actual rotation after 10000 cycles $\varphi_{t=10000}^{\text{av}}$ is for smaller D_r. This means, that the serviceability might get lost although the structure rotates back during subsequent cyclic loading.

5.3.1.4. Influence of the cyclic loading direction

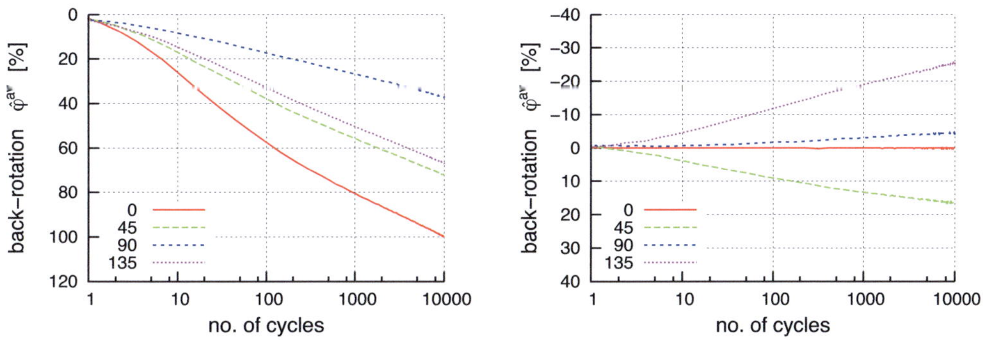

(a) Normalised rotation on x-direction

(b) Normalised rotation in y-direction

Figure 5.8.: Rotations of the foundation in the 3d simulation with varying directions of the cyclic loads, see Figure 4.3

The development of the accumulated back-rotation for different directions α of the cyclic load package is shown in Figure 4.3; Figure 5.8a shows the the rotation parallel, while Figure 5.8b shows the rotation perpendicular to the direction of the extreme load; compare Figure 4.3. The rotation ψ^{av} in Figure 5.8b is normalised, viz.

$$\hat{\psi}^{\text{av}} = \frac{\psi^{\text{av}}}{\Delta\varphi_{\text{ref}}^{\text{av}}} \tag{5.4}$$

As one would expect, the largest back-rotation measured in the direction of the extreme loading is obtained for $\alpha = 0°$, the smallest for $\alpha = 90°$ and an intermediate for $\alpha = 45°$ and $\alpha = 135°$. This is, however, different for $\hat{\psi}^{\text{av}}$. The foundation rotates in the direction of the average value for $\alpha = 45°$ while it rotates against it for $\alpha = 135°$. No rotation is obtained for a loading in the direction $\alpha = 0°$, which confirms the suitability of plane strain model. In order to interpret the result, it is favourable to decompose of the loads in components with respect to the reference coordinate system, shown in Figure 4.3; see Table 5.5.

Table 5.5.: Decomposed load components according to coordinate system of Figure 4.3

load	direction α	x-comp.	y-comp.
F_{ext}	$0°$	F_{ext}	0.0
F	$0°$	F	0.0
	$45°$	$F/\sqrt{2}$	$F/\sqrt{2}$
	$90°$	0.0	F
	$135°$	$-F/\sqrt{2}$	$F/\sqrt{2}$

Since the cyclic load amplitude F^{ampl} has the largest influence on the amount of accumulated back-rotation (Figure 5.6), it is plausible that a load in direction $\alpha = 90°$ causes the smallest, in $\alpha = 0°$ the largest and in $\alpha = 45°$ and $\alpha = 135°$ an intermediate back-rotation. The different signs of the x-components for $\alpha = 45°$ and $\alpha = 135°$ are responsible for the different rotations $\hat{\psi}^{\text{av}}$ of the foundation perpendicular to the extreme loading. Furthermore, the extreme load has caused a change of the incremental stiffness of the soil under the foundation. It is stiffer on the compression side than it is on the uplift side.

5.3.2. Settlement behaviour

Experiences with cyclically loaded shallow foundations have revealed that significant settlement may occur even in dense sand. They are uncritical for the stability of a structure, but are relevant for the serviceability. The settlement is a composition of (re-)compaction, i.e. densification, and of squeezing under constant volume, i.e. when the density of the soil under the foundation remains constant. Both can be hardly discriminate, except for foundations on very dense sand and void ratios close to e_d[7].

The development of the settlement is independent of the average value F^{av} of the cyclic load, as shown in Figure 5.9. Neither the shape nor the residual rotation after 10000 reversals depends on F^{av}. The settlement increase linear with the logarithm of the number of cycles.

Figure 5.10 shows the influence of the cyclic load amplitude F^{ampl} on the accumulated settlement. One can see that $\hat{s}^{\text{av}}$ increases with increasing amplitude. The rate of settlement $\dot{\hat{s}}^{\text{av}}$ is almost proportional to F^{ampl}.

[7] e_d is a lower boundary of maximum density of a soil under exclusion of grain crushing

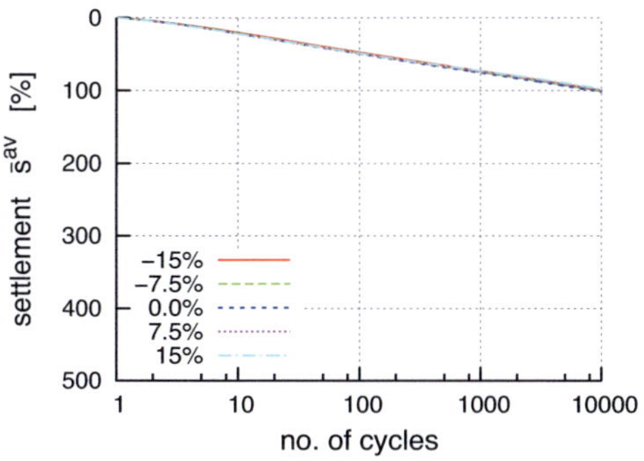

Figure 5.9.: Influence of F^{av} on $\hat{s}^{\text{av}}$ for $D_r = 85\%$ and $F^{\text{ampl}} = 10\%$

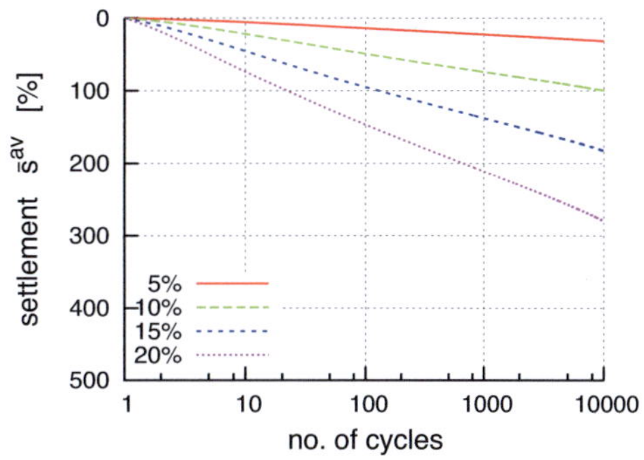

Figure 5.10.: Influence of F^{ampl} on $\hat{s}^{\text{av}}$ for $D_r = 85\%$ and $F^{\text{av}} = +7\%$

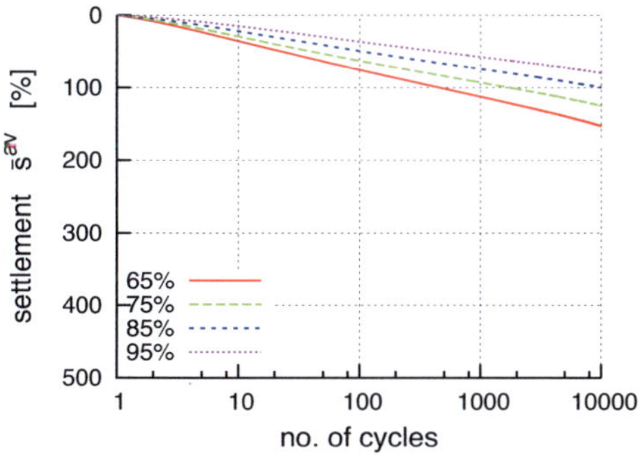

Figure 5.11.: Influence of D_r on $\hat{s}^{\text{av}}$ for $F^{\text{av}} = +7\%$ and $F^{\text{ampl}} = 10\%$

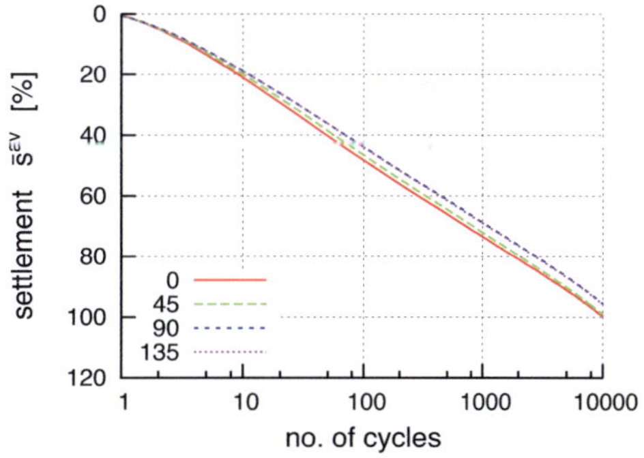

Figure 5.12.: Influence of the loading direction α on $\hat{s}^{\text{av}}$ for $D_r = 85\%$, $F^{\text{av}} = +7\%$ and $F^{\text{ampl}} = 10\%$

The same proportionality, but less distinct, can be seen in Figure 5.11 in which the influence of the initial density on the development of the accumulated settlement is shown. It is reasonable, that $\hat{s}^{\text{av}}$ increases with decreasing D_r. The amount of compaction of a loose soil is obviously larger than of a dense soil. Hence the settlement for decreasing initial relative density has to be larger.

In Figure 5.12 is shown the influence of the loading direction on $\hat{s}^{\text{av}}$. Although the magnitude of the amplitude influences the accumulated settlement, the change of the loading direction and hence the change of the magnitude of the cyclic load components, as listed in Table 5.5, has almost no influence on $\hat{s}^{\text{av}}$. From this result one can conclude, that not the direction of a cyclic load, but solely its amplitude governs the average settlement of a foundation.

5.3.3. Summary

An evaluation of the back-rotation after 10000 cycles $\hat{\varphi}^{\text{av}}_{t=10000}$ of all computed variations listed in Table 5.3 is presented in Figure 5.13. A similar evaluation is shown in Figure 5.13 of the accumulated settlement $\hat{s}^{\text{av}}_{t=10000}$. The numerical results for $F^{\text{ampl}} = 15\%$ are not plotted in Figure 5.13 for reasons of clarity. The corresponding curves lie in-between $F^{\text{ampl}} = 10\%$ and 20% and have a similar shape as the other curves.

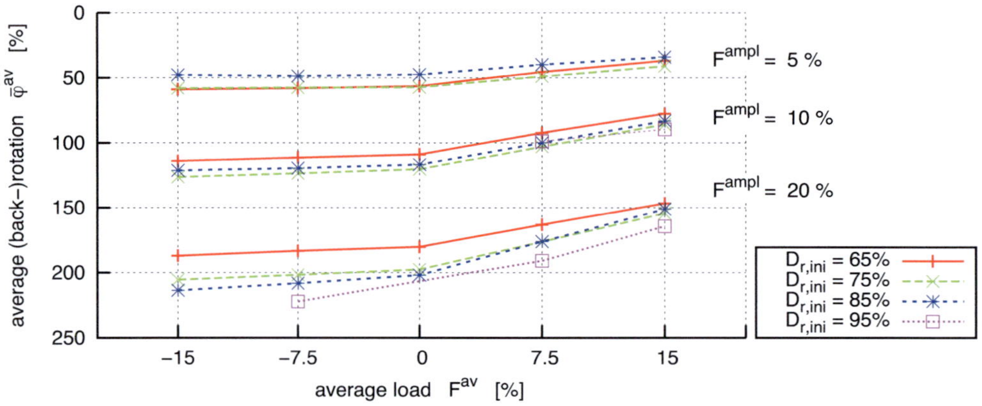

Figure 5.13.: Accumulated back-rotation after 10000 cycles for different amplitudes, average values and initial densities

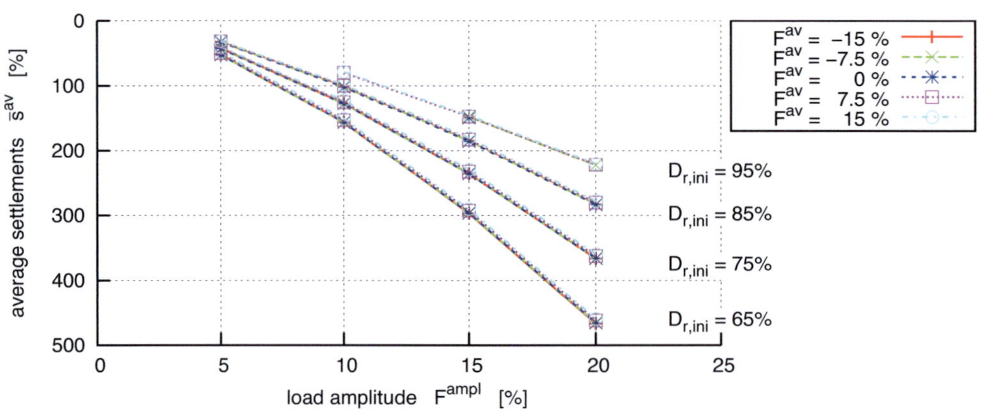

Figure 5.14.: Accumulated settlement after 10000 cycles for different amplitudes, average values and initial densities

One can see that mainly the cyclic load amplitude governs the amount of accumulated back-rotation, while the influence of the average value and initial density is secondary. Negative values of F^{av} have even no influence at all. The relative density D_r shifts the curves vertically, but does not affect the shape. Besides of $F^{\mathrm{ampl}} = 5\%$, it is apparent from Figure 5.13, that the amount of normalised back-rotation is higher with dense sand. I.e., a back-rotation is favoured for foundations on dense sand.

From Figure 5.14 becomes evident, that the amount of accumulated settlement significantly increases with decreasing initial relative density and increasing cyclic load amplitude. If one compares Figure 5.13 with Figure 5.14, it becomes apparent, that a doubling of the accumulated back-rotation may cause triplication of the settlement. Hence, settlement can be an important point to consider in the design.

5.3.4. Influence of drainage

Since the initiation to this study arose from a research project dealing with foundation for offshore wind turbines, the influence of water also considered in the analyses of the stabilisation behaviour. Pore water generally inhibits volume changes of the soil during (un-)loading and shearing, depending on the permeability of the soil, the governing drainage path and the rate of loading. Instead of compaction or dilatancy, pore pressure and suction, respectively, is generated. Following the principle of effective stresses, the soil can thus *soften* or *harden*. The observed behaviour of the foundation on dry sand can be both, intensified or extenuated, depending on the loading time and permeability of the soil. Settlement will increase if pore pressure is generated or decrease if significant suction is mobilised; latter in general only on the uplift side.

Figure 5.15 shows results of simulations of model tests on saturated sand. Arrangement, dimensions, initial conditions as well as loading conditions correspond to the tests presented in Section 4. The simulations were done again in prototype scale. Differing from the tests on dry sand, a load period of 1/20 Hz for the extreme loading and 1/5 Hz for the subsequent cyclic symmetrical load packages was assumed. Furthermore, resting periods were neither allowed between extreme and cyclic load step nor between the load sequences[8]. The applied load F_{ext} corresponds to the design load; i.e. half of the actually in the model test applied load.

The results presented in Figure 5.15 are normalised according to Equation (4.1). Figure 5.15a shows that the main characteristics observed for dry sand can be rediscovered for saturated conditions; the back-rotation and the *hardening*. Somewhat surprising is, that the maximum rotation $\bar{\varphi}_{\max}$ decreases while the mount of accumulated rotation $\Delta\bar{\varphi}_{t=500}$ increases with decreasing

[8] refer to 4.3 and Figure 2.6

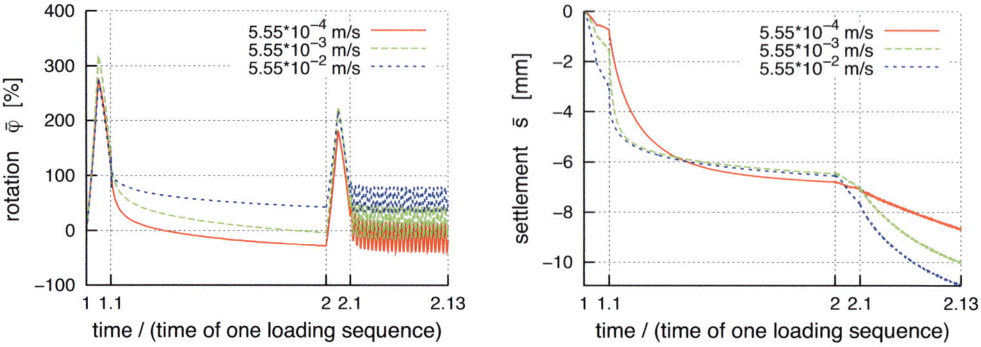

(a) Normalised rotation

(b) Normalised settlement

Figure 5.15.: Numerical predictions of the model test on saturated sand and different permeability with *hca* and *hypo+igs* (first cyclic load step) and *hypo+igs* (second cyclic load step)

permeability k of the soil. Corresponding to that, the average settlement of the foundation also decrease during with decreasing permeability, which is shown in Figure 5.15. Similar results are obtained for the foundation presented in Chapter 7. This behaviour can be also observed for other simulations employing other constitutive models. The reason for this behaviour is the influence of suction on the uplift side, which will be discussed later in Chapter 6.

However, the results of the simulations have to be taken with care. The calculated behaviour is qualitatively plausible, but the ratio of the back-rotation is debatable. A complete recovery of the foundation seems to be unrealistic, although large symmetric cyclic loads were applied on the foundation. There are several reasons by which this strange behaviour in simulations can be explained, viz.

- **Constitutive model:** The number of control cycles, which have to be computed implicitly in order to update the accumulation rate for *hca*, has only a minor influence on the computed total amount of accumulated stresses and strains. This has been studied in detail by WICHTMANN [150] and could be confirmed by the author. However, all studies were done so far for dry or completely saturated sand under undrained conditions wherein a volume conservation is presumed. This is, however, not the case for the simulations shown in Figure 5.15 in which simultaneously volumetric and deviatoric deformations as well as stress relaxation occurs.

  Besides of the performed number of control cycles, the internal elastic stiffness of the high cycle accumulation model could have an influence on the result (Wichtmann, personal communication).

- **FE-model:** The convergence study with respect to size, density and type of element as well as increment length has been done for dry conditions. Since transient formulation requires different elements, the convergence study include coupled simulations.

5.4. Comparison of different constitutive models

Since all simulations so far were done by means of *hypo+igs*, a brief review of the stabilisation behaviour with respect to the chosen constitutive model is presented in this section. It has been shown recently by PRADA [122], that SaniSand, a representative of elasto-plastic models, is to some extent suitable to describe both, quasi static and cyclic loading of soil samples. A back-calculation of laboratory element tests on Toyoura sand, shows a good accordance, see Figure 3.10. Since a parameter determination is out of the scope of this work and not necessary for a comparison, the following simulations were done with parameters for Toyoura sand. Model dimensions, as well as initial and loading conditions were chosen in accordance to the simulations presented in Section 5.2. The comparison was done by means of implicit models only; i.e. *hca* was not employed, although this would have been possible. In order to minimise the accumulated numerical errors due to the integration procedure, only the first 25 cycles of the cyclic load steps were computed. The results are presented in Figure 5.16.

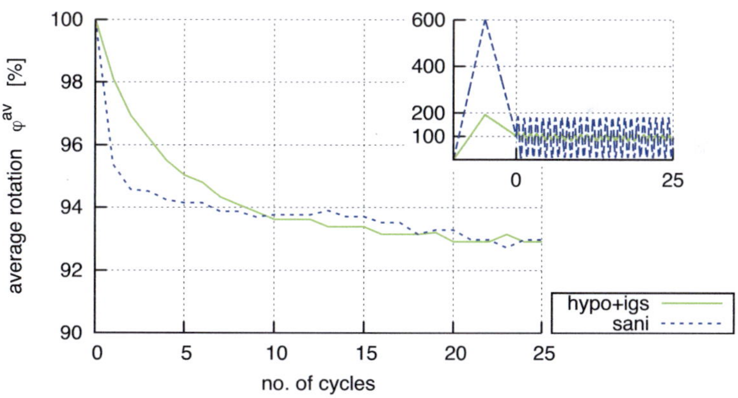

Figure 5.16.: Normalised rotation computed with *hypo+igs* and SaniSand

A back-rotation is obtained with both constitutive models. The fact that both models end up with the same normalised amount of back-rotation is accidental. If more cycles have been computed, the curves would intersect, since the rotation curve predicted with *hypo+igs* is more inclined. The jagged shape of the shown curves is due to the evaluation method; the simulations itself are numerically stable.

Although the rotational response to the cyclic loading agrees well with respect to the average back-rotation, the overall response of both models differs significantly. The maximum rotation $\bar{\varphi}_{\mathrm{max,Sani}}$ is almost three times larger than the hypoplastic one. Also the strain amplitude in the cyclic load step is noticeably larger with SaniSand. This can be caused by the lack of a lower bound in the elasto-platic model, as already discussed in Chapter 3.2. The initial relative density is $D_r = 95\%$, i.e. it is close to the maximum density e_d. It seems that those high densities are captured less accurately than by *hypo+igs*. This disadvantage was also figured out by PRADA [122].

Anyway, from the comparison it becomes apparent, that the back-rotation can be qualitatively predicted with advanced constitutive models and is not an artificial artefact of *hypo-igs* which accidently agrees to observations in model tests. This is a further proof of the suitability of the employed model.

6. Discussion of the stabilisation behaviour

A paradoxical behaviour of a cyclically loaded shallow foundation on sand has been presented in Chapter 4 and Chapter 5. Based on experiences from element tests, one would expect that an initially tilted foundation rotates further on, if subjected to cyclic loading.

One-dimensional cyclic tensile tests on iron rods show a gradual softening, which is independent on the order of the applied load cycles with varying amplitude; e.g. MINER [103]. The term softening is understood as a continuous decrease of the stiffness modulus E. Similar observations have been made with one-dimensional cyclic compression tests on concrete cubes. These findings are the basis of today's calculation procedures for the dimensioning of cyclically loaded structural elements, such as beams, shells and plates made of steel or concrete. The softening of the material is considered by so-called *Wöhler-lines* or *S-N-curves*. They are a function of the load amplitude and the number of applied load reversals; see also Chapter 2.

WICHTMANN [150] could show by means of cyclic stress-controlled triaxial tests on dry sand samples, that granular materials obey also the Miner-rule. That means, the sum of accumulated average strains in stress-controlled cyclic tests is independent on the order of the applied load amplitudes; e.g. NIEMUNIS ET AL. [114]. However, contrary to element test conditions, our foundation is a boundary value problem. Hence, the considered back-rotational behaviour can hardly be compared with an element test.

A literature review revealed, that similar observations of back-rotating foundations have not been reported so far. [1] Hence, a discussion of this novel and unexpected behaviour is necessary. In order to describe this phenomena, the terms *self-healing* and *stabilisation* have already been employed in earlier chapters; but an explanation has not been given yet. This will be done in this chapter by means of an identification of governing mechanisms.

6.1. (Differential) Settlement

Unequal settlement of a static or cyclically loaded shallow foundation is in general accompanied by an unsymmetrical stress distribution in the contact zone between subsoil and structure.

[1] One exception is a large scale offshore model test done in frame of the construction of the *Oosterschelde* storm flood barrier. A detailed description and a numerical back-calculation is presented by STURM ET AL. [139, 140].

But also inhomogeneous subsoil under a foundation may cause unequal settlement even if the stresses in the soil-structure interface are uniform and symmetrical. This is the case if the subsoil under the foundation has varying density or consists of inclined layers of soft soil with varying thickness. Besides of unsymmetrical stress distribution and varying soil properties, also the foundation geometry may lead to differential settlement, which is in particular relevant for tall buildings; e.g. BURLAND ET AL. [30]).

The settlement of cyclically loaded shallow foundations has been studied intensively by several authors. Three different (loading) cases can be distingusihed:

Cyclic vertical loading Almost all studies focused on vertical load reversals. This is a rather simple problem, since the behaviour of a boundary value problem is symmetric. The settlement is even and a tilting of the foundation can be neglected. Different settlement rules were proposed for this load case, which base either on laboratory element tests (e.g. DIYALJEE AND RAYMOND [54], MALLWITZ AND HOLZLÖHNER [97], HOLZLÖHNER [71] and SAWICKI ET AL. [127, 128, 129]), or on results of model tests (e.g HOLZLÖHNER [70] and HETTLER [69]). The accumulated average settlement of a cyclically and vertically loaded shallow foundation generally increases linearly with the logarithm of the number of load reversals. The accumulation rate is a function of load amplitude, average load and soil density (HOLZLÖHNER [70]), as well as of foundation geometry and unit weight of the soil (HETTLER [69]).

Cyclic horizontal loading Studies on foundations, subjected to (elevated) cyclic horizontal loads can hardly be found in the literature. Some research was done by means of model tests for the design of projected offshore foundations; e.g. LAMBE ET AL. [94], ALLARD [3] and ANDERSEN [16]. A systematical study of different parameters has not been done so far.

Arbitrary cyclic vertical and horizontal loading A study on arbitrary cyclic loading and its influence on differential settlement, cannot be found in the literature.

In order to study the mechanism of (differential) settlement and a subsequent back-rotation, it is convenient to distinguish between two different components of settlement: compaction (Figure 6.1a) and isochoric squeezing (Figure 6.1b). Although a strict distinction is difficult, numerical simulations with the finite element method allow to separate both components and to evaluate them individually, if strains are considered instead.

The compaction, or densification, is due to a change of the void ratio $\Delta e = e - e_{\text{ini}}$ only, while grains can be considered incompressible and grain crushing can be excluded for the prevailing stresses. The volumetric strain volumetric strain is defined as

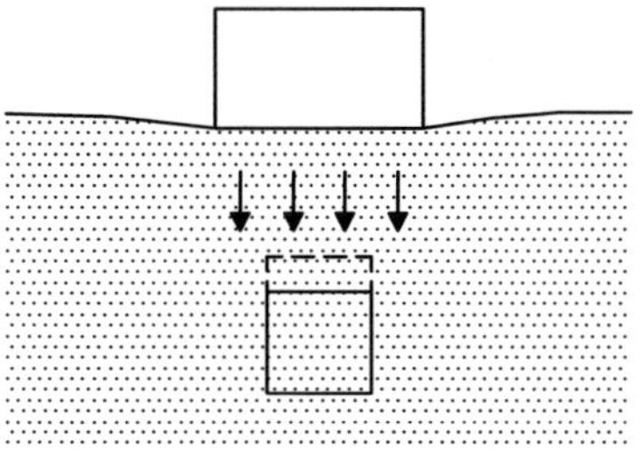

(a) Compaction

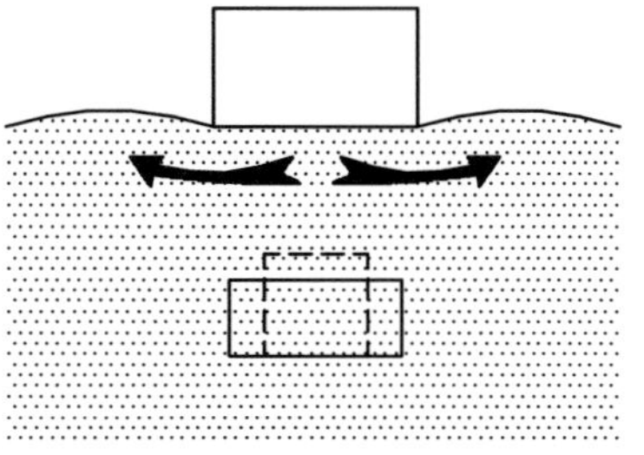

(b) Squeezing

Figure 6.1.: Two mechanism causing settlement of a foundation; KOLYMBAS [89]

$$\varepsilon_v = \text{tr}\,\varepsilon = \varepsilon_1 + \varepsilon_2 + \varepsilon_3 = \frac{\Delta e}{1 + e_{\text{ini}}} \qquad \text{with} \qquad \varepsilon_i = \ln\left(\frac{l}{l_0}\right)_i \tag{6.1}$$

Alternatively, Δe can be normalised with $(e_{\max} - e_{\min})$. Therein, $e_{\max}$ and $e_{\min}$ are upper and lower limits of void ratios, determined by means of standard index tests for granular materials; e.g. DIN 18126 [48] or ASTM 4253 and 4254 [10, 9]). This normalisation corresponds to a change of the relative density ΔD_r.

$$\begin{aligned} \Delta D_r &= \frac{\Delta e}{e_{\max} - e_{\min}} = \frac{e_{\text{ini}} - e}{e_{\max} - e_{\min}} \\ &= \frac{(e_{\max} - e_{\max}) + (e_{\text{ini}} - e)}{e_{\max} - e_{\min}} = \frac{(e_{\max} - e) - (e_{\max} - e_{\text{ini}})}{e_{\max} - e_{\min}} \\ &= D_r - D_{r,\text{ini}} \end{aligned} \tag{6.2}$$

However, ΔD_r could be a misleading representation of a densification. Since the initial relative density prior to the installation amounts already 95%, D_r will exceed 100% after installation due to the additional vertical pressure. Thus, for the subsequent discussion, ε_v will be used instead. An alternative representation to ΔD_r could be the pressure corrected relative density r_e; HERLE [67].

The amount of isochorically squeezed out material can be quantified with the deviatoric strain

$$\varepsilon_q = \|\mathbf{e}\| = \sqrt{e_1^2 + e_2^2 + e_3^2} \qquad \text{with} \qquad e_i = \varepsilon_i - \frac{\varepsilon_v}{3} \tag{6.3}$$

The norm $\|\mathbf{e}\|$ can be considered as a *shear-deformation-variable* which describes deformations of a soil element, except of volume change. Equation (6.3) can be simplified for plane strain conditions to

$$\varepsilon_q = \sqrt{\left(\frac{2}{3}\right)} \cdot \sqrt{\varepsilon_1^2 + \varepsilon_2^2 - \varepsilon_1 \cdot \varepsilon_2} \quad . \tag{6.4}$$

Both strain measures ε_v and ε_q are scalar field variables. Hence they are suitable to present in contour plots. In Figure 6.2 and Figure 6.3 are shown the results of the back-calculated model test, which was presented in Section 5.2. The initial extreme loading, which should represent a severe storm or a freak wave, is applied in positive x-direction. Hence the compression side is on the right. In order to distinguish the influences of the the different loading steps on the strain measures, differential volumetric and deviatoric strains are presented. That means, the strains after unloading from the extreme load are reduced by the initial strains after installation. This is shown in Figure 6.2a and Figure 6.3a. Likewise the strains at the end of the cyclic load step, i.e. after 500 cycles, are reduced by the residual strains after unloading from the extreme load, shown in Figure 6.2b and Figure 6.3b).

The initial extreme load F_{ext} causes an asymmetrical stress distribution in the soil-structure interface. Hence the settlement on the compression, i.e. right, side is larger and the foundations tilts. This can be seen for both strain measures, volumetric (Figure 6.2a) and deviatoric strains (Figure 6.3a). Since the soil is initially already very dense with $D_r = 95\%$, the deviatoric strains are larger than the volumetric ones. However, despite of the high density, a loosening of the subsoil due to horizontal sliding and shearing can hardly be observed. A dilatancy of the soil can be seen only close to the outer skirts on the uplift side (left side).

In the subsequent cyclic loading step, the compaction and isochoric squeezing of the soil under the foundation increases on both sides. However, from Figure 6.2b and Figure 6.3b becomes apparent, that the amounts of $\Delta\varepsilon_v$ and $\Delta\varepsilon_q$ are larger on the former uplift side. That means that the foundation rotates to the left during cyclic loading. In conjunction with the strains shown in Figure 6.2a and Figure 6.3a, the foundation actually rotates back. The total tilt decreases and the structure straightens up again.

Like for the extreme loading before, the deviatoric strains dominate. But the differences between compression and uplift sides are not that pronounced as they were for the extreme loading. The reason for that is, that the average settlement of the foundation due to cyclic loading is significantly larger than due to the extreme loading. This became already apparent from Figure 5.3b.

From Figure 6.2 and 6.3 can be seen that the back-rotation of a shallow foundation is accompanied by settlement. Based on the explanations presented in the next section, we will see, that settlement is necessary but not sufficient for self-healing (back-rotation).

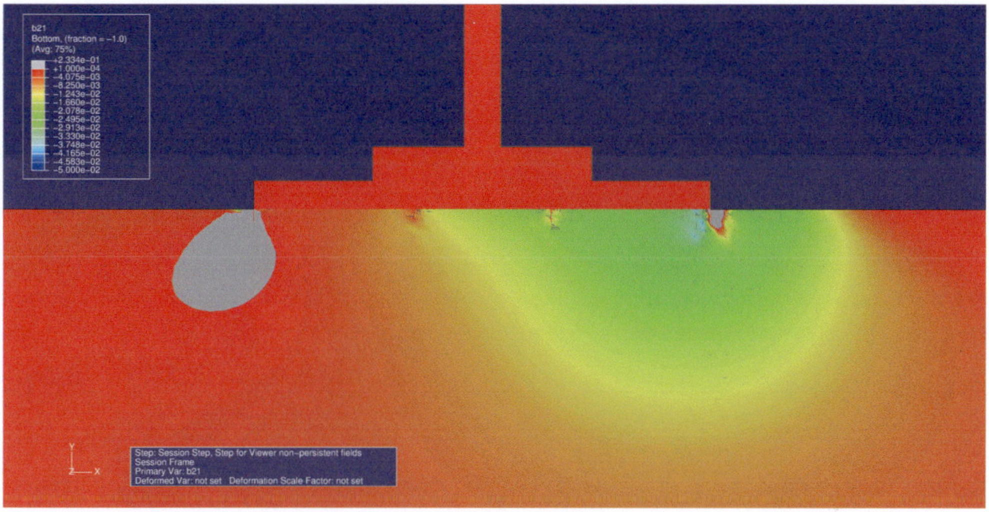

(a) Differential volumetric strains after extreme loading ($\varepsilon_{v,\varphi_0} - \varepsilon_{v,\text{ini}}$)

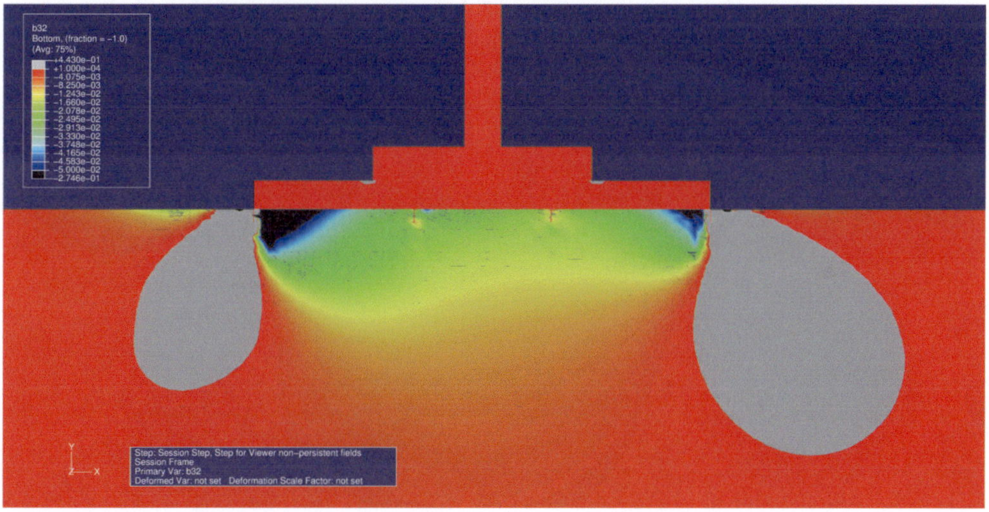

(b) Differential volumetric strains after 500 load reversal ($\varepsilon_{v,\varphi_{500}} - \varepsilon_{v,\varphi_0}$)

Figure 6.2.: Differential volumetric strains $\Delta\varepsilon_v$. Areas of compaction are coloured, while areas of loosening are grey. The range of compaction reaches from red (no settlement) over green to blue and black (large settlement).

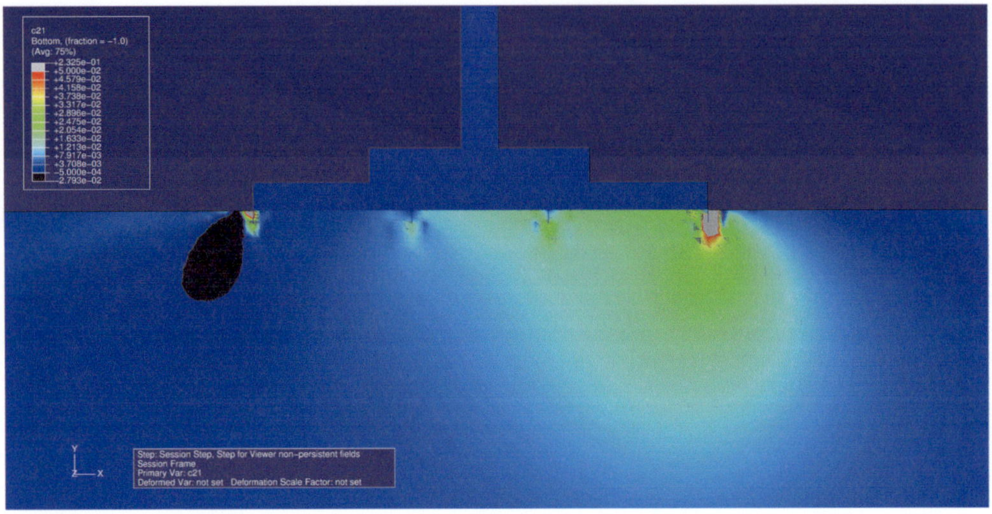

(a) Differential deviatoric strains after extreme loading ($\varepsilon_{q,\varphi_0} - \varepsilon_{q,\text{ini}}$)

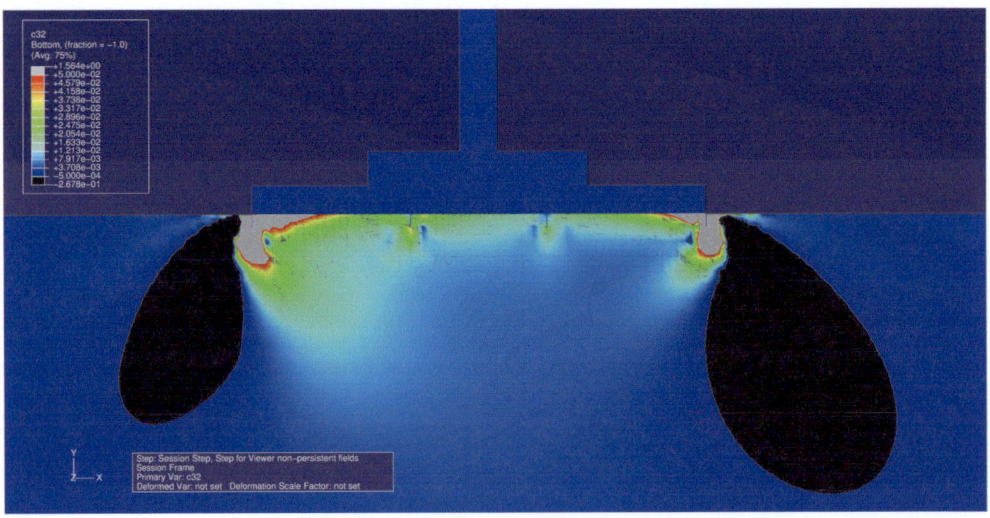

(b) Differential deviatoric strains after 500 load reversals ($\varepsilon_{q,\varphi_{500}} - \varepsilon_{q,\varphi_0}$)

Figure 6.3.: Differential deviatoric strains $\Delta\varepsilon_q$. Areas of increased shear deformations are coloured, while areas of decreased shear deformations are black. The range of shear deformations reaches from blue over green to red and grey.

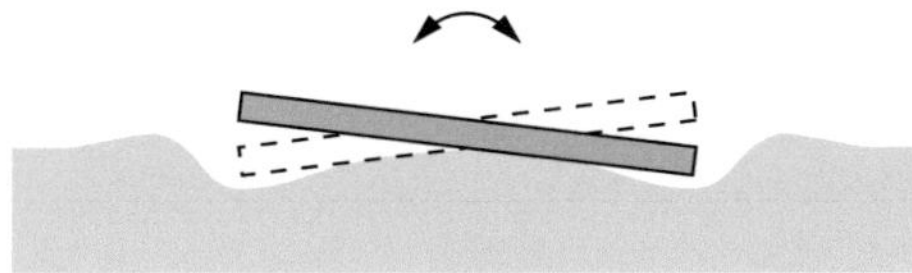

Figure 6.4.: A riding foundation during cyclic loading

Besides of the back-rotation, another typical behaviour becomes evident from Figure 6.2 and 6.3: the so-called riding. Both, differential volumetric and deviatoric strains have a symmetrical component; i.e. the settlement of the soil on the uplift and compression sides are larger than in the centre. The foundation *seesaws* from side to side over a soil column in the symmetry axis of the foundation, which does not experience significant settlement (Figure 6.4). With every load reversal, the soil under the current compression side will be additionally densified and squeezed. I.e. the rotational amplitude increases. During an uplift phase, the danger of erosion under the plate for offshore conditions increases. This phenomenon as well as its consequences are followed up in Chapter 7.

6.2. Two explanatory models

6.2.1. Soil mechanical approach

WICHTMANN [150] has performed an extensive laboratory testing program of cyclic stress-controlled triaxial tests on dry sand. One purpose of these tests was the development of a cyclic flow rule for granular materials under general loading conditions. He varied initial density D_r, confining (average) pressure $p^{\mathrm{av}} = (\sigma_1 + 2 \cdot \sigma_3)/3$, deviatoric stress $q^{\mathrm{av}} = \sigma_1 - \sigma_3$ and cyclic load amplitude q^{ampl}. He employed the so-called ACCUMULATED STRAIN RATE $\dot{\varepsilon}^{\mathrm{acc}}$ in order to quantify the influence of the different parameters, which is defined viz.

$$\dot{\varepsilon}^{\mathrm{acc}} = \frac{\mathrm{d}\varepsilon^{\mathrm{acc}}}{\mathrm{d}N} \quad \text{with} \quad \varepsilon^{\mathrm{acc}} = \sqrt{(\varepsilon_P^{\mathrm{acc}})^2 + (\varepsilon_Q^{\mathrm{acc}})^2} \tag{6.5}$$

$$\text{with} \quad \varepsilon_P = \frac{1}{\sqrt{3}}(\varepsilon_1 + 2 \cdot \varepsilon_3) \quad \text{and} \quad \varepsilon_Q = \sqrt{\frac{2}{3}}(\varepsilon_1 - \varepsilon_3)$$

with $\varepsilon^{\mathrm{acc}}$ denoting the sum of the the accumulated (i.e. plastic) volumetric ε_v and deviatoric ε_q strains. WICHTMAN [150] could show that $\dot{\varepsilon}^{\mathrm{acc}}$ *decreases* for

- *increasing D_r* and constant p^{av} and $\eta^{\mathrm{av}} := \dfrac{q^{\mathrm{av}}}{p^{\mathrm{av}}}$;
- *increasing p^{av}* and constant η^{av} and D_r;

- *decreasing* η^{av} and constant p^{av} and D_r.

These findings can be employed for a first simplified approach of the back-rotation. Let us consider two representative soil elements under the foundation; one on the compression and one on the uplift side. Their state prior to the extreme load can be assumed to be equal.

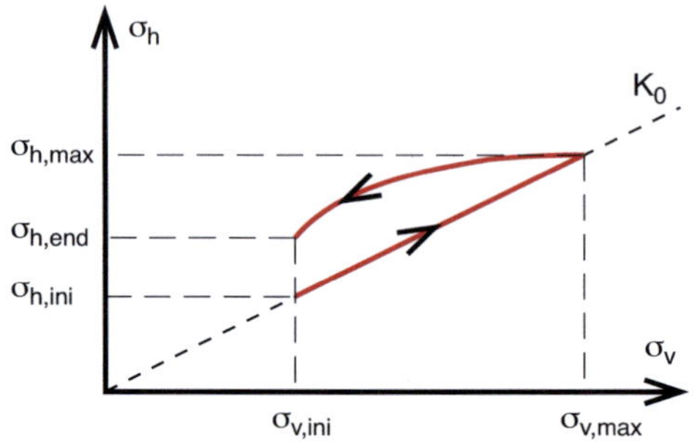

Figure 6.5.: Development of horizontal stress for oedometric loading and unloading

BAUER [20, 21] has performed one dimensional compression tests on dry sand with a *soft-oedometer*. This device allows the measurement of horizontal stresses σ_h. The samples were prepared to a K_0 stress state. They were than subjected to a temporary increase of the vertical stress σ_v. Bauer could show, that the horizontal stress σ_h was larger after the temporary loading than initially, although, σ_v was equal to its initial value; see Figure 6.5. I.e. that the grains in the sample can be considered to be more jammed after this loading, since $\sigma_{h,\text{end}} > \sigma_{h,\text{ini}}$ but $\sigma_{v,\text{end}} = \sigma_{v,\text{ini}}$.

Based on this finding, one may assume that the stress ratio η^{av}_{comp} of a representative soil element on the compression side will be smaller and $D_{r,\text{comp}}$ and p^{av}_{comp} larger than initially.[2] Analogous, $D_{r,\text{up}}$ and p^{av}_{up} are smaller on the uplift side and η^{av}_{up} is larger than initially. The soil experience a *temporary induced alignment* of stresses and porosity matrix. I.e. the force-chains as well as the grain matrix have preferred orientation.

In the subsequent cyclic loading step, both representative soil elements under the foundation are subjected to vertical cyclic loading. The stress amplitude σ_v^{ampl} is about equal on compression and uplift side. Following the results by Wichtmann, the average accumulated strain – and hence its rate – will be larger on the uplift side, since the loading conditions is equal for both elements, but the initial state is different. As a consequence, the foundation will rotate back.

[2] Actually, the representative soil element is additionally subjected to a small shear stress τ which can be, however, for the present neglected. τ is small, since the application point of the horizontal load, which acts on the superstructure, is relatively high. Hence, the vertical forces resulting from the moment, dominate the loading of the representative soil elements.

If the average horizontal force F^{av} acting on the structure increases during cyclic loading, the average stress state of both soil elements will change too. The vertical stress on the compression side $\sigma_{v,\mathrm{comp}}$ will be larger, while $\sigma_{v,\mathrm{up}}$ on the uplift side will be smaller. Hence, the elements tend to different accumulated stress states. The accumulation rate of the element on the uplift side will be smaller, than for a symmetric loading. The back-rotation rate will be also smaller, which agrees with the numerical results.

The loading conditions of the considered boundary value problem are more complex. The behaviour of the soil elements under the foundation can hardly be compared to either oedometric or triaxial conditions. The boundary conditions of the considered elements are variable; vertical and horizontal stresses and strains will change during cyclic loading due to a redistribution of stresses as well as due to densification and shearing of the element. Thus, the state and the loading conditions of the soil elements resemble different kinds of laboratory tests. Despite of the complexity, the presented model may helps to yield a mechanism of back-rotation. Furthermore, it does not contradict the aforementioned Miner rule.

6.2.2. Physical approach

Almost all current constitutive models for the description of mechanical behaviour were derived from observations made with laboratory element tests, model tests and in-situ measurements. They are mainly empirical and have hardly a physical background. Although some models have been proved suitable to predict even complex soil-structure interactions for various static and cyclic loading conditions, none of them is universally valid, but rather restricted to a special range of application.

In general, the models assume quasi-static deformations of *non-transient*, i.e. permanent, soil elements. Other aspects of granular materials such as rate-dependence are often neglected. More complex processes such as demixing or phase transitions from solid to liquid states and back are completely disregarded. On the other hand, especially the prediction of liquefaction got into focus of recent research activities. The risk of landslides during intense rain or earthquakes affects many people. Often users of current constitutive models claim to predict the safety against liquefaction. But actually, they can at best predict whether liquefaction occurs under the presumed conditions. But these model are not able to determine the safety against it, since they do neither incorporate descriptions for the mechanical behaviour of fluids nor transition states between solids and fluids.

The approach presented in the following, is an attempt to employ established physical principles in order to describe soil mechanical phenomena. The idea is not new. Several authors of constitutive models have tried to prove, that their models adhere to thermodynamic principles.

Examples are the *hyperplasticity*, proposed byHOULSBY [75], and the *endochronic theory* proposed by VALANIS [146, 148]. The authors followed thereby two different approaches: they either tried to derive the mathematical formulations of their already existing constitutive models from thermodynamic relations or they showed that their models do not contradict thermodynamic principles.

The *granular soild hydrodynamics* (GSH) model, proposed by Gudehus and Liu, bases completely on physical principles and claims to overcome the aforementioned restrictions of existing models. Independent works from GUDEHUS [62] and JIANG AND LIU [85] employed a so-called *granular temperature* T_{g} in order to describe the *seismic activity* of a grain skeleton. T_{g} is a scalar field variable, which should represent the amount of transferred seismic energy into the absolute temperature T. The seismic energy arises from local transpositions of a grains causing a release of energy which dissipates wave-like into the surrounding grain skeleton. The concept of a granular temperature is not new, but has already been proposed earlier by HAFF [64] and JENKINS [84]. Gudehus and Liu have modified and enhanced the concept and came up with a new physically sound model for the description of the mechanical behaviour of granular soils.

The GSH model bases on the concept of the total free energy F. The model assumes that the total free energy is composed of several energy components such as gravitational E_g, elastic E_e, kinematical E_k, seismic E_s and thermal energy E_t, viz.

$$F = E_g + E_e + E_k + E_s - E_t \tag{6.6}$$

Gudehus could show, that only three energy components are relevant for the description of quasi static deformations at constant temperature, which is the case for most engineering problems, such as the analysed foundation and the self-healing behaviour. The free energy F is an integral over the volume of specific gravitational, elastic and seismic energy densities, viz.

$$F = \int_V (w_g + w_e + w_s)\ \mathrm{d}V \tag{6.7}$$

Therein, w_g denotes the potential energy of structure and grains. w_e represents the elastic energy of the jammed grain skeleton. And w_s describes the specific seismic energy of the jittering of the grains.

A foundation experiences through its superstructure a range of *seismic* excitations due to the cyclic loading, which can have a wide frequency spectrum. They are transferred via the foundation into the subsoil and propagate there like waves. A fraction of the propagated seismic energy dissipates into total temperature T via the granular temperature T_{g}. It has a similar effect on the grains as the total temperature on a molecule. The granular temperature is a kind of *catalyser* which aids to minimise the free energy F of a system.

Let us consider again the two representative elements under the foundation. As stated in the previous section, the states of both elements are different; the density is different and the grains are jammed on both sides. The cyclic forces which acting on the foundation during normal operation induced a seismic excitation of the grain skeleton in both elements. The *mobility* of the grains on the uplift is obviously larger than on the compression side, due to the different relative densities. Hence the stresses relax and the density increases faster, too. That means, the amount of dissipated free energy, and hence performed work, is larger on the uplift side. Consequently, the foundation will rotate back.

The GSH model confirms by means of physical principles the observations made on dry sand in cyclic triaxial tests performed by WICHTMANN [150], as presented in the previous section. It hence justifies the empirical approach employed to describe the mechanism causing the back-rotation. Although, GSH is currently still in the development, JIANG AND LIU [85] could already show, that both, the hypoplastcity model as well as the high cycle accumulation model are special cases of the granular solid hydrodynamics model. This justifies, besides of the successful back-caclulation of the model test, the suitability of both models for the simulation of this novel behaviour; namely the back-rotation during cyclic loading.

6.3. Pore water and drainage

This thesis arose from a research project on the behaviour of cyclically loaded foundations of offshore wind turbines. This raises the question about the influence of pore water and drainage conditions on the back-rotational behaviour. The laboratory tests mentioned in Section 6.2 were perfomed with dry sand. The results, however, can be also employed, at least qualitatively, for saturated sand with and without drainage.

Instead of a (re-)densification, changes of the pore water pressure u are generated by undrained cyclic loading. The rate of accumulation $\dot{u}$ is larger for lower relative density D_r. I.e. the average mean pressure $p'^{\mathrm{av}} = p^{\mathrm{av}} - u^{\mathrm{acc}}$ decreases and $\eta^{\mathrm{av}} = q^{\mathrm{av}} / p'^{\mathrm{av}}$ increases. Following the results of Wichtmann, the accumulated average strain rate $\dot{\varepsilon}^{\mathrm{acc}}$ will hence be larger on the uplift side, and the foundation will rotate back, too.

Sand, however, has a considerable permeability, and drainage has to be taken into account. This has been done for the simulations presented in Section 5.3.4. The results showed, that the foundation rotates back even for partially drained conditions. Numerical simulations with low permeable soils accumulated faster pore water pressure, which cause an increase of accumulated pore water pressure. That means, that the effective stresses on the compression side decreases faster too. However, the unloading and horizontal shearing of the representative soil element on the uplift will accumulate negative pore water pressure, i.e. *suction*. The positive effect of suction can be seen in Figure 5.15. With increasing suction decrease the maximum rotation during extreme loading and increases the rate of the back-rotation, because the suction has the effect of a additional restoring force on the foundation.

6.4. A visco-hypoplastic description of the self-healing

The influence of load reversals on the average accumulated state of a soil element can be described as relaxation, respectively, creep, which depends on the loading conditions as shown in Figure 6.6. A sand sample in a triaxial testing device, subjected to stress cycles with constant amplitude and open drainage, will accumulate strains, which can be considered as a kind of creep; Figure 6.6a. Analogous yield strain cycles with constant amplitude to a stress relaxation; Figure 6.6b. But also mixed loading cases are possible in which creep and relaxation occurs.

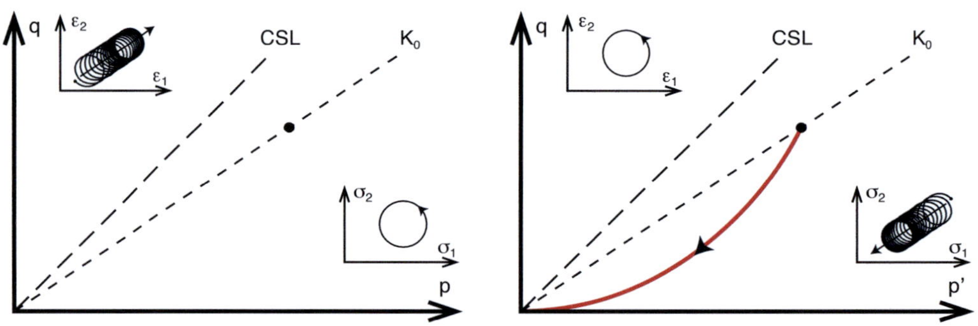

(a) Stress-controlled test on a drained sample (b) Strain-controlled test on an undrained sample

Figure 6.6.: Cyclic triaxial test on sand for different loading and drainage conditions

The terms *creep* and *relaxation* are generally used to describe viscous soil properties. It is well known, that a clay sample, subjected to constant stresses will creep. On the other hand, the stresses acting on a sample under constant volume and shape will relax. Hence, it should be possible to simulate the back-rotational behaviour of the foundation with the visco-hypoplastic constitutive model. The relaxation rate of viscous soils decrease linearly with the logarithm of time, which is similar to the observed and calculated behaviour of the model foundation during cyclic loading. The rate of viscous materials is a function of the current stress state as well as of the relative stress-dependent density, generally denoted as over-consolidation ratio (OCR); Equation (3.10) in Chapter 3.

Based on theses similarities, GUDEHUS has proposed a so-called *seismo-hypoplasticity* for the description of the behaviour of granular materials subjected to cyclic loading [62, 63]. The constitutive model utilised a visco-hypoplastic approach in order to describe creep and relaxation [61]. However, the development has been given up in favour of the above presented GSH model, in which several concepts of *seismo-hypoplasticity* can be rediscovered.

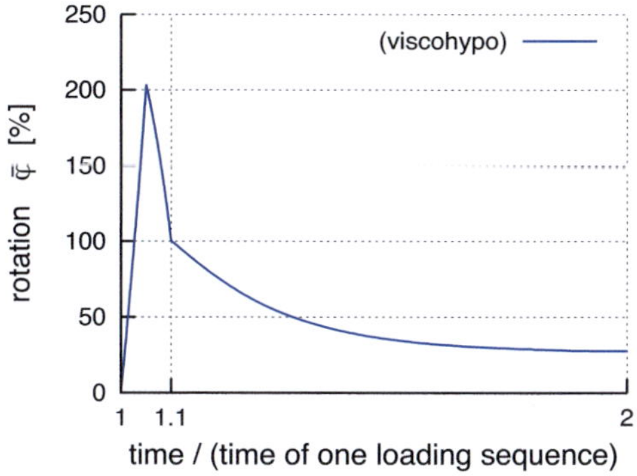

Figure 6.7.: Qualitative simulation of the model test with a visco-hypoplastic model

In Figure 6.7 is shown a numerical simulation of the rotational behaviour of our shallow foundation, employing the visco-hypoplastic model proposed by NIEMUNIS [109, 112], presented in Chapter 3. The same finite element mesh was used as shown in Figure 5.2. The constitutive parameters were taken from WIENBROER [77], who has back-calculated qualitatively static creep and relaxation tests on sand, performed by DI BENEDETTO [53]. The parameters are listed in Table 6.1.

Table 6.1.: Visco-hypoplastic parameters used for the simulation of the shallow foundation

φ_c	λ	κ	e_{100}	$\dot{\gamma}$	I_v	β	m_T	m_R	$R_{\max}$	β_χ	χ
35°	0.03	0.012	0.49	$1 \cdot 10^{-7}$	0.055	0.95	5.0	5.0	$1 \cdot 10^{-4}$	0.1	1.0

The soil prior to the extreme loading was assumed to be normal consolidated. The extreme load was applied rapidly, which explains the low ratio of maximum rotation $\varepsilon_{\max}$ and residual rotation after unloading ε_0. In a subsequent step, the soil under the foundation could creep or relax, respectively, under the dead weight of the structure. This boundary conditions correspond to cyclic symmetrical horizontal loads, as done in the model tests.

One can see, that the foundation rotates back, which can be explained again by the state of two representative soil elements. The element on the compression side after the extreme load has a higher OCR than the element on the uplift side. Hence, the creep rate on the latter side during static loading is larger than on the compression side. Thus the foundation rotates back.

6.5. Concluding remarks

The rate of settlement of a cyclically loaded shallow foundation depends on the state of the subsoil. The rate decreases for decreasing average deviatoric stress and increasing mean pressure and density. The state of the soil under a tilted foundation, due to an extreme load event, is inhomogeneous. I.e. the deviatoric stress is larger and the density and mean pressure are smaller on the uplift side than on the compression side. Hence, the foundation rotates back during subsequent cyclic loading.

Since the vertical alignment of a structure is necessary for its serviceability, it is appropriate to denote the back-rotation *self-healing*. The term *stabilisation* is justifiable to describe the mechanism, since the rate of back-rotation decreases with increasing number of load reversals. A stable state is characterised by a (local) minimum of the free energy, whereas unstable states can be identified by an acceleration of the system.

From the presented model tests and simulations, the impression may arise, that every shallow foundation rotates back under the considered loading conditions. This is, however, not true. E.g. a tall building, which has a high centre of gravity and is founded on a narrow foundation. This building will be unable to rotate back from a tilted position due to an extreme loading. The eccentricity of the centre of gravity is too large and the stress distribution at rest is significantly unsymmetrical.

There are several reason why this particular behaviour has not been observed earlier. First of all, neither suitable constitutive models were available nor sufficient hardware performance. Also the analytical and semi-analytical models presented in Chapter 2 often oversimplified the boundary value problem of shallow foundation. For example SEED AND IDRISS[132] suggested to consider a representative soil element under a horizontal cyclic loaded foundation as an element in a direct simple shear test. No distinction were made between compression and uplift side and vertical cyclic loads were neglected. This simplification was justifiable in the past for offshore structures with smaller overturning moments, but are not suitable for OWTs anymore.

7. FE-simulations of a substructure founded on multiple plates

In Chapter 5, a structure founded on a large single plate was analysed. This type of foundation has been successfully employed to found *Offshore Wind Turbines* (OWT) in water depths of up to 10 m, e.g. NYSTED [117] and HORNSREV [72]. However, most projected offshore wind parks in the German sector of the North and Baltic seas are located in regions with water depths between 20 and 40 m. Under these conditions, this foundation geometry is uneconomical (for details refer to Appendix A). Hence, other geometries should be considered in order to preserve the advantages of shallow foundations compared with deep foundations, as discussed in Chapter 2.

One possible geometry of a shallow foundation is proposed in this chapter: a substructure in shape of a cross resting on the subsoil via four plates attached at the outer side of the cross, shown in Figure 7.1. All components, including shaft and plates are made of reinforced concrete. The idea to found an OWT on a cross-shaped foundation, with and without plates, originates from a proposal of the German construction company Ed. Züblin AG. The shape of a cross for the substructure has been adopted in this work for reasons of simplicity. But the results of the simulations are not restricted to the cross. They are also valid for other geometries as long as the substructure

- is founded on four feet;
- has a significant large weight compared to the total weight of the structure;
- and that the centre of gravity is lower than the diameter of the foundation.

The geometry and dimensions of the analysed foundation were chosen based on experiences gained from a numerical study of a cross-shaped foundation with plates, done in frame of a research project financed by the Ed. Züblin AG; STURM [138]. It is shown in Figure 7.1 and will be denoted in the following as *cross-foundation*.

Although, this is a new approach for the foundation of OWTs, and many properties could be discussed here, main focus in this chapter is spent on the study of the behaviour during cyclic loading with respect to the stabilisation; that means the back-rotation and settlement. Therein, two aspects are of main interest:

1. Behaves the cross-foundation similar to the foundation analysed in Chapter 5?
2. How large is the influence of the specific structural elements of the cross-foundation on the stabilisation behaviour?

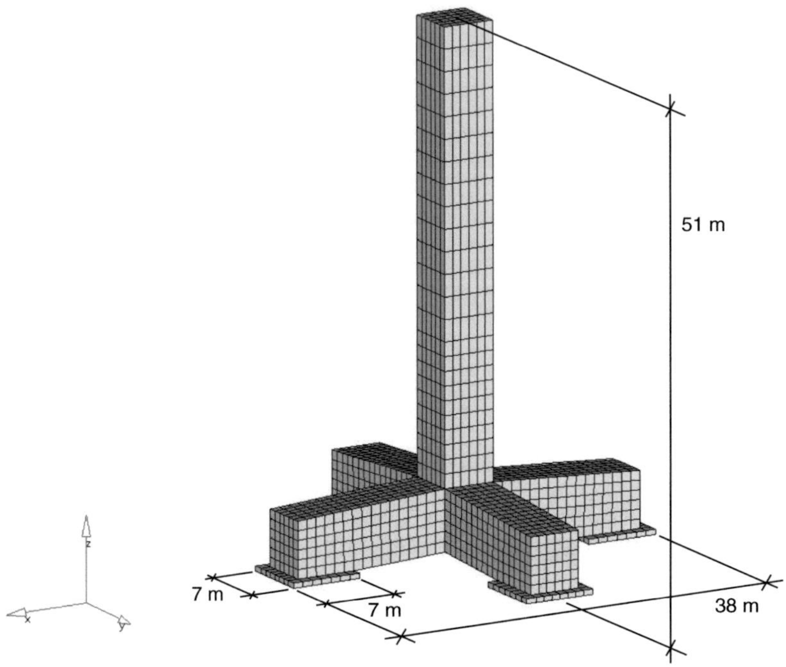

Figure 7.1.: Dimensions of the analysed substructure for an offshore wind turbine in shape of a cross resting via four feet on the subsoil. Cross, feet (plates) and lower shaft are made of reinforced concrete. The hollow arms and shaft are are with ballast, such as gravel or stones.

The cross-foundation has some advantages compared to the foundation analysed in Chapter 4 and Chapter 5.

- Conventional shallow foundations consisting of a large single plate tend to *ride* if subjected to cyclic loading; see Chapter 6 for details. Since the riding is accompanied by a gradually alternating compaction and squeezing of the subsoil on compression and uplift side, an increased rotational amplitude of the foundation is often observed under continuous cyclic loading with constant stress amplitude. The plate will be undermined and erosion occurs. Contrary to that, the cross-foundation will not *ride* since only the outer plates are resting on the subsoil, but not the centre.

- The arms of the cross are hollow and are filled with ballast after the foundation is installed on site. Hence the structure prefabricated onshore can be rather light which is favourable for the manufacturing and transportation.
- The construction of a cross-shaped substructure made of reenforced concrete is somewhat more costly than of a single plate. But it is still simple to build and much cheaper than other geometries, such as e.g. a ring-shaped substructure or a steel frame.
- The transportation and installation is an important issue, especially if one considers that the projected offshore wind parks should comprise of up to several hundred OWTs. The shaft provides a good handle and the cantilever can be relatively short; the crane can operate close to the shaft betwixt two arms. Züblin has designed and built a vessel with a notch in shape of the cross in its hull. With it, transportation and installation can be done by one unit.
- The use of four feet has some important advantages over three feet; a static over-determination leads to a permanent redistribution of the stress distribution under the foundation plates during loading. In contrast to a statically determined structure the foundation with four feet is less sensitive on lopsided loads, typically for offshore conditions.
- The substructure can be lighter than a conventional shallow foundation consisting of one plate only. The static vertical load including ballast, has to be just as high as necessary to prevent the foundation from sliding. Apart from that, the above mentioned properties have be nevertheless complied.

7.1. Structural concept

Early offshore platforms for oil- and gas-production, founded on shallow foundations, were equipped with no or relatively short skirts; see Chapter 2. One example of a skirt-less foundation is the *Frigg CDP1* platform, which was installed in 1975; LACASSE ET AL. [93, 92]. Already after the first severe storm, piping and erosion below the platform was observed. Although the overall stability was not endangered – according to official information – the platform was already decommissioned in 1990 after only 15 years of operation. All following platforms of similar type were equipped with skirts.

In order to assess the influence of skirts, one has to distinguish between short skirts, which serve as scour protection only, and longer skirts which provide additionally an increase of the bearing capacity due to an embedment depth. The terms short and long are referred to the diameter of the foundation plate; see Chapter 2.

7.1.1. Platforms without skirts

The design is rather simple, since skirts have not to be considered. Hence, the proof of safety against sliding according to established standards can be employed to determine the minimum required dead weight, i.e. the vertical load V_{stat} at rest.

$$V_{\text{stat}} \geq \eta_{\text{slide}} \cdot \frac{1}{\tan \delta} \cdot H_{\text{ext}} \tag{7.1}$$

Equation (7.1) assumes that the foundation is not embedded into the soil. Therein, δ is the friction angle between concrete plate and subsoil, and η_{slide} the safety factor against sliding. The required minimum diameter of the foundation, i.e. the length of an arm of the cross, can be determined by means of the proof of safety against overturning, viz.

$$d \geq \begin{cases} \eta_{\text{ot}} \cdot M_{\text{ext}} \cdot V_{\text{stat}}^{-1} \cdot 2 + a & \text{(7.2a)} \\ \eta_{\text{ot}} \cdot M_{\text{ext}} \cdot V_{\text{stat}}^{-1} \cdot 2\sqrt{2} + a & \text{(7.2b)} \end{cases}$$

with a being the side length of the quadratic plates and η_{ot} the safety factor. In Equation (7.2) is assumed, that the centre of rotation coincides with the geometrical centre of a plate, hence the diameter d is reduced by the length a. It is also assumed, that the weight of the foundation is distributed equally over the cross. The edge length a can be determined by means of the proof of safety against bearing failure, viz.

$$M_{\text{ext}} \leq \begin{cases} \frac{1}{4} V_{\text{stat}} \cdot \frac{1}{2} \cdot \frac{1}{2} d + \left(R_1 \cdot \frac{1}{\eta_f} - \frac{1}{4} V_{\text{stat}} \right) \cdot \frac{1}{2} d & \text{(7.3a)} \\ 2\frac{1}{4} V_{\text{stat}} \cdot \frac{1}{2} \cdot \frac{1}{2\sqrt{2}} d + \left(R_2 \cdot \frac{1}{\eta_f} - 2\frac{1}{4} V_{\text{stat}} \right) \cdot \frac{1}{2\sqrt{2}} d & \text{(7.3b)} \end{cases}$$

with R_1 and R_2 representing the resistance against failure of one plate (Equation (7.3a)) and two plates (Equation (7.3b)), respectively, which can be determined according to established national and international standards. The need to distinguish two cases arises from the foundation geometry and the loading directions, discussed in the following. The values R_1 and R_2 are functions of the edge length a. The described procedure is an iterative process, which has to be recalculated several times in order to optimise the geometry.

The equations [7.2] and [7.3] distinguish between two different loading directions: parallel to an arm of the cross ([7.2a], [7.3a]) and along the bisectrix of two arms ([7.2b], [7.3b]). The differences can be seen in the Figures 7.2 and 7.3. While it is evident that Equation (7.2b) governs the required diameter, it is not that obvious for the edge length a. Equation (7.3) considers, besides the resistance factors R_i, an additional restoring force due to the eccentricity of V_{stat} referred to the geometrical centre of the dominating plate. Since it is reasonable to assume that the plates perpendicular to a loading along an arm do not uplift, only a quarter of the total vertical load is considered in Equation (7.3a). Analytical and numerical calculations show, that this loading direction is the governing case.

In the design of similar structures was assumed that the horizontal load is carried only by the plates perpendicular to the loading direction while the increased vertical loads, due to the overturning moment, by the plates in loading direction. This approach is, in the opinion of the author, not justifiable since it is kinematically inconsistent. A reaction force can be mobilised only if the subjected structural element deforms.[1] Since the axial stiffness of the superstructure is always significantly larger than the shear stiffness[2] of the soil, all four feet carry almost the same amount of the horizontal load. Hence, the resistance factors R_i in Equation (7.3) has to withstand the increased vertical load due to the overturning moment as well as on the horizontal load. For a rough estimation, it is justifiable to assume homogenous stress distribution in the interface.

Although skirts have not been considered in this section, it should be noticed, that one should include this structural element. It has been shown, that skirts are an effective protection against scour, e.g. WHITEHOUSE [149]. They can be rather short compared to the diameter of the foundation plate.

7.1.2. Platforms with skirts

If platforms are equipped with sufficiently long skirts which are effectual with respect to the bearing capacity, a different design approach than the one presented in the previous section has to be chosen. The embedded soil columns between the skirts may increase the effective weight of the structure. Also the resistance against sliding increases due to the mobilised passive earth pressure behind the skirts. The same holds for the determination of the bearing capacity. Additionally, the skirts on the uplift side may mobilise a suction under offshore conditions due to negative pore water pressure, which depends on the skirt length, diameter of the plate and permeability of the soil.

The proposed shallow foundation with longer skirts is a hybrid of a *Gravity Base Structure* (GBS) and a (suction-) bucket foundation; refer to Chapter 2 for details. The latter foundation type has been installed at the *Draupner* and Sleipner oil- and gas-fields on the Norwegian shelf in North sea; BYE ET AL. [32] and ERBRICH AND TJELTA [56]. The ratio between skirt length

[1] in this case, the plates have to slide in order to mobilise friction

[2] actually one has to compare the stiffness of the structure with the friction angle δ under consideration of the dimensions of the structural elements

to diameter of bucket amounts 0.5 for these platforms. These foundations carry a significant portion of horizontal and moment loading by means of mobilised suction on the uplift side. The superstructure of both platforms is made of a steel similar to traditional jacket platforms founded on piles.

7.2. FE-model

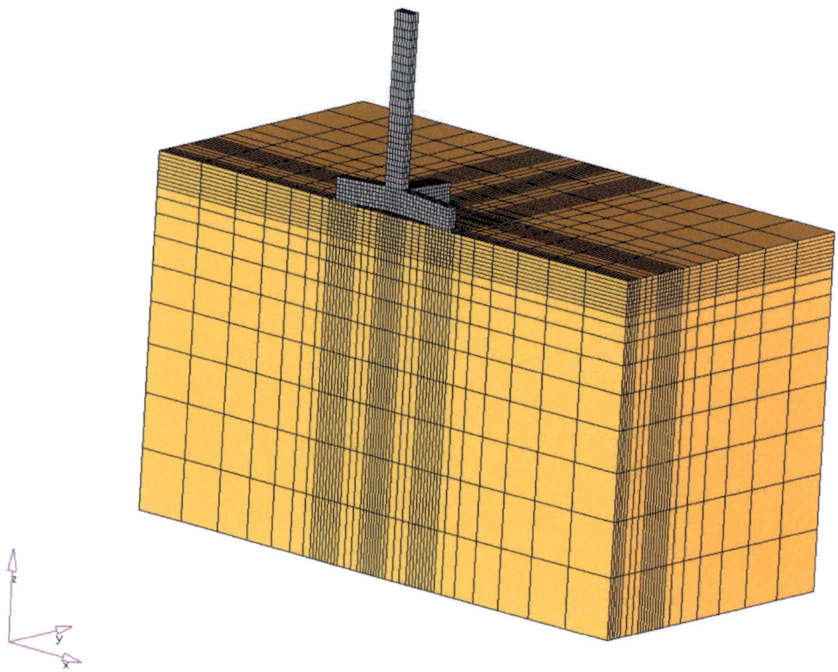

Figure 7.2.: A half 3d FE-model of the cross-foundation with loading along an arm

Figures 7.2 and 7.3 show two 3d FE-models of the cross-foundation. Both models utilise the symmetry axes of the foundation, hence only half of the foundation had to be modelled. The loading direction in Figure 7.2 is along an arm of the cross and in Figure 7.3 along the bisectrix of two arms. Due to the rectangular shape of the foundation, all finite elements could be modelled rectangular, too. Hence, the element quality can be considered as good with respect to its characteristic values[3], except of the aspect ratio which is larger than five for some elements. However, these elements are located sufficiently far away from the plates of the foundation.[4]

[3]compare Section 5.1 for further details on mesh quality

[4]the distance is approximately three times the plate diameter

One could have used a node-*biasing* at the vertical boundaries. This would have caused, however, distorted elements with acute and obtuse angles. As already discussed in Section 5.1, those elements may cause numerical difficulties in simulations with transient pore water flow.

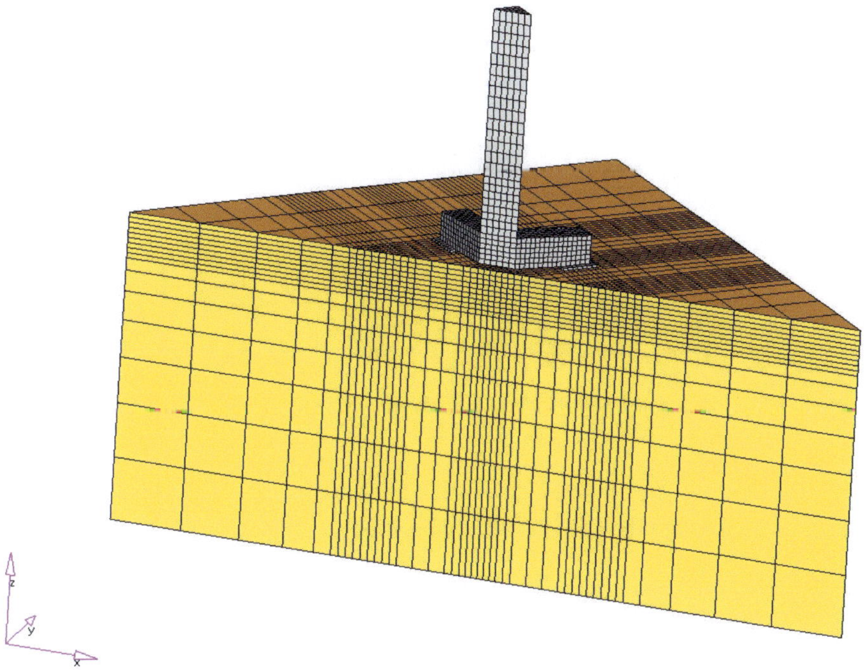

Figure 7.3.: A half 3d FE-model of the cross-foundation with loading along the bisectrix of two arms

Neither the circular boundary of the test-cell, describe in Section 4, nor the dimension of the model tests were adopted. Instead, a relatively large portion of the half-space was modelled, in order to guarantee that the numerical results are not affected by the boundary conditions of the soil. The modelled soil block has a length and width of 120 m and a depth of 90 m. Hence, it is $\approx$ 13 times deeper and $\approx$ 17 times wider than the edge length a of a plate. The dimensions of the foundation are shown in Figure 7.1. They were determined by means of the above presented analytical approach as well as by numerical simulations. Of course, also for this model an extensive convergence study with respect to element type and size as well as degree of discretisation has been done; STURM [138].

Figure 7.4 shows a simplified 2d FE-model of the cross-foundation with skirts under the plates. The model utilises the same symmetry axis already used in the model shown in Figure 7.2. The arm and hence plates perpendicular to the model plane are not considered. One could

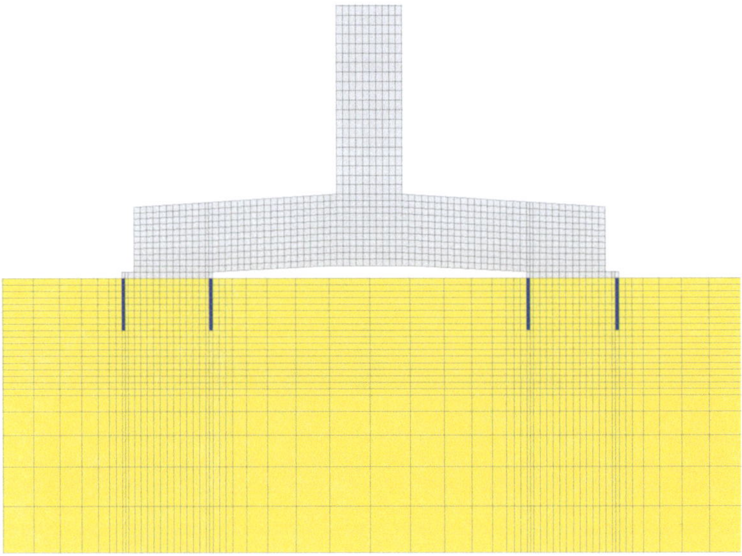

Figure 7.4.: 2d FE-model of the cross-foundation with loading along an arm

have introduced springs attached beneath the shaft in the symmetry axes of the plane strain model in order to approximate the effect of the statical indeterminacy of the structure. But a reliable estimation of the spring stiffness is hardly feasible.

Since only two of the four plates are considered in the 2d FE-model, weight and horizontal load are halved, since both are carried by all four plates equally. The size of the overturning moment, however, remains unchanged since the section modulus is almost independent of the plates perpendicular to the modelled plane.

$$I_y = \frac{1}{3} \cdot a^4 + 2 \cdot \left(\frac{d}{2} - \frac{a}{2}\right) \cdot a \tag{7.4}$$

Equation (7.4) is also valid for a loading along the bisectrix of two arms. I.e. I_y is independent on the loading direction.

The plane strain model reduces the 3d problem of the cross-foundation to a bisected pipe, as illustrated in Figure 7.5. Thus, the spatial deformation mechanism is likewise reduced to a plane strain problem. This simplification is accompanied by a decrease of the bearing capacity, hence the settlement of the plates are overestimated. Furthermore, the omitted plates perpendicular to the model plane circumvent a redistribution of the stresses between the four interfaces. Both limitation of the 2d model should be kept in mind by the evaluation of the stabilisation behaviour during cyclic loading presented in the subsequent sections. The differences on the rotational and settlement behaviour between 2d and 3d model has been analysed in detail by STURM [138].

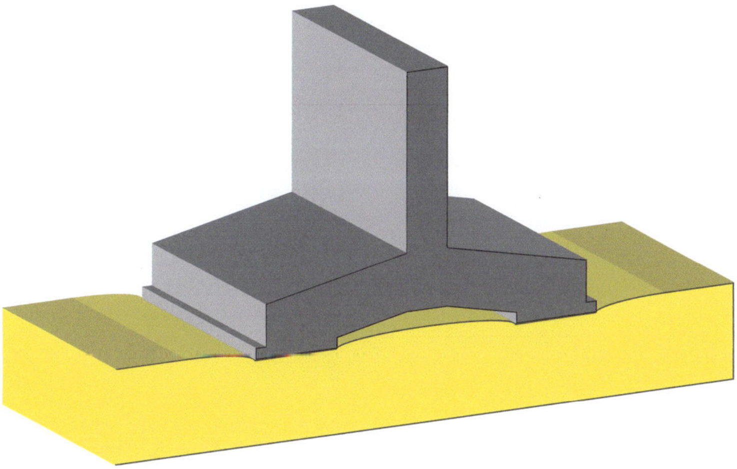

Figure 7.5.: Extrusion of the simplified 2d model of the cross-foundation. The simplified 2d-model of the foundation is comparable to bisected pipe lying upside down on the subsoil.

All models incorporate interfaces both between plates and soil as well as between skirts and soil. A soil-structure friction angle of $\delta = \varphi_c = 32.5°$ was assumed. In the simulations with transient pore water flow, soft elastic and high permeable elements were arranged between the underside of the plates and the topside of the soil. They take into account the vertical in- and outflow of pore water into a possible gap between soil and plate, which is relevant only for very short skirts with significant asymmetric vertical displacements. These elements allow also the computation of suction, i.e. negative pore water pressure, and avoid inadmissible tension, i.e. negative effective stresses, assuming pressure being positive, in the soil.

7.2.1. Prediction of model tests

The simulations presented in this section should reveal the behaviour of the cross foundation subjected to the same loading conditions as employed in the model tests presented in Chapter 4. For this, the 3d model without skirts, as shown in Figure 7.4, was used. Like for the simulations shown in Chapter 5, the analyses was done in prototype dimensions. The extreme load F^{ext} as well as the cyclic load amplitude F^{ampl} correspond to the design load; refer to Appendix A for details. I.e. F^{ext} – and F^{ampl} – in the numerical simulations was only half of the actually applied load in the test.

The computed rotational response of the cross foundation is shown in Figure 7.6, the corresponding settlement in Figure 7.7. Other than for the plate-foundation analysed in Chapter 5, only the hypoplastic model with intergranular strain was employed in these simulations. The high cycle accumulation model could not be used due to restrictions of the employed hardware as well as a missing suitable Fortran compiler. Some special features provided only by some compilers are required to run the user-routine of the high cycle accumulation model in conjunction with the FE-prgram ABAQUS. In order to minimise accumulated numerical errors due to the implicit calculation procedure of the hypoplastic model with intergranular strain, only the first two test sequences with 25 cycles each were computed.

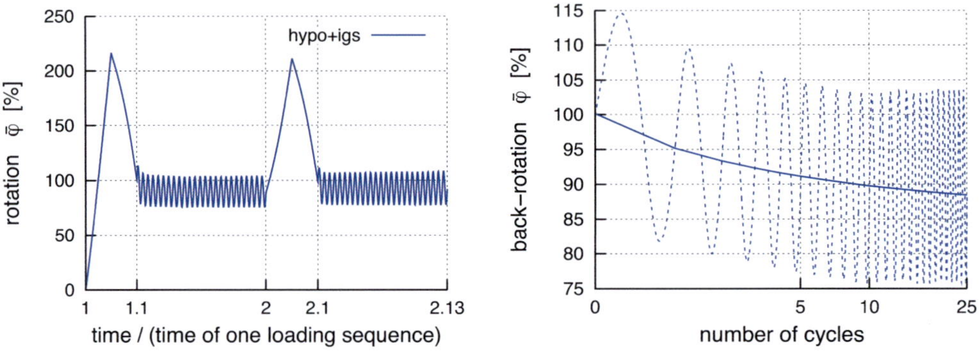

(a) Rotation for the first two test sequences

(b) Development of the back-rotation in the first cyclic load step

Figure 7.6.: Calculated rotational response of the cross foundation under model test conditions

Figure 7.6a reveals, that the cross-foundation behaves similar as the plate-foundation. All relevant properties observed in the model tests are reproduced. Due to the limited number of computed load cycles, the amount of back-rotation is relatively small. As a result, also the *hardening* is less distinct. The maximum rotation for the second extreme loading is almost equal to the rotation for the first loading. The difference between both is only 5%; from 215% in the first sequence to 210% in the second sequence. Same is valid for the amount of back-rotation.

While Figure 7.6a implies that the back-rotation can only be seen for the first 2 to 3 cycles, a detailed plot of the rotation versus the logarithm of the number of cycles, as shown in Figure 7.6b, reveals, that the back-rotation is a continuous process which still goes on at the 25th cycle. An asymptotic approach to a linearly decreasing $\bar{\varphi}^{\text{av}}$, as seen for the plate-foundation, seems to be also possible for the cross-foundation, but cannot be said for sure due to the small number of computed load cycles.

The predicted average settlement is presented in Figure 7.7. It is split into its components on the uplift and compression sides, respectively. Like in Figure 5.3b, the computed settlement is significantly larger than in the model tests. Also the dilatancy behaviour can hardly be reproduced. The predicted uplift on the lee side is even somewhat smaller compared to the simulations with the plate-foundation.

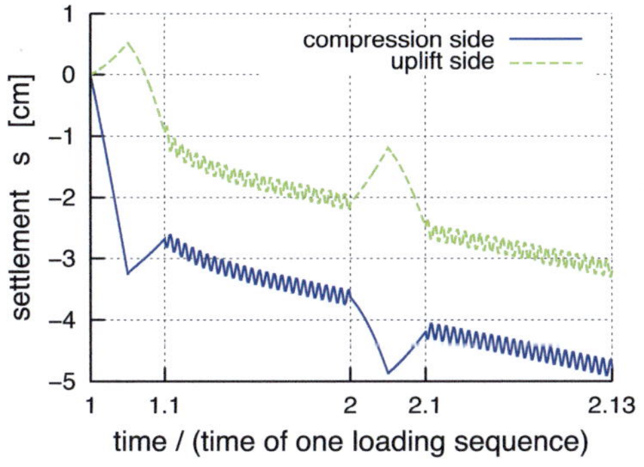

Figure 7.7.: Calculated settlement response of the cross foundation under model test conditions for the first two test sequences on the uplift and compression side of the cross-foundation

The differences between the model tests of the plate-foundation and the simulations with the cross-foundation may have several reasons, which should be kept in mind during the evaluation of the numerical analyses which will be presented in the following sections.

- The reduced *hardening* behaviour, i.e. the low increase of the incremental stiffness, could be a result of the small number of computed load reversals. The achieved amount of (re-)compaction is obviously less for 25 cycles than for 500. Hence, also the predicted back-rotation has to be smaller.
- The simplifications of the 2d FE-model has a direct influence on the predicted stabilisation behaviour: $\bar{\varphi}_{\max}$ can be expected to be smaller and $\Delta\bar{\varphi}^{\text{av}}$ to be larger for the 3d FE-model than for the 2d FE-model, due to the above described restrictions; STURM [138].
- The stresses in the soil-structure interfaces are at rest significantly larger for the cross-foundation as for the plate. But since the initial relative density of $D_r = 95\%$ is similar in both simulations, the stress-corrected relative density r_e is larger, hence the predicted incremental stiffness of the soil has to be smaller. As seen in the simulations of the plate-foundation on sand with $D_r < 95\%$ and constant stresses of the structure, the amount of back-rotation decreases with decreasing density. This behaviour is equivalent to increased stresses but equal density and implies, that the volumetric accumulation rate $\dot{\varepsilon}_v^{\text{acc}}$ has to be larger for the cross-foundation than for the plate. Consequently, the settlement on the compression and the uplift side are not as different any more as seen for the plate-foundation.

7.2.2. Parameter study

In the following a numerical parameter study on the influence of the back-rotational behaviour of the cross-foundation is presented. Based on the structural differences between cross- and plate-foundation, two aspects are of main interest in this study:

1. the influence of the (elastic) stiffness of the cross-shaped superstructure connecting the four plates;
2. the influence of the skirts length, i.e. the embedment depth of the foundation.

Table 7.1.: Values of the varied parameters in the numerical parameter study on the back-rotational behaviour of the cross-foundation.

Quantity	**Variations**
bulk modulus $E_s/E_{s,ref}$	10 %, 100 %, 1000 %
skirt length h_i	0 m, 1 m, 2 m, 4 m

Table 7.1 lists the performed values of the varied parameters. The study is done by means of the *hypoplastic model with intergranular strain* (hypo+igs). The constitutive parameters shown in Table 5.1 were employed in the simulations. Like for the predicted behaviour under model test conditions, presented in the previous section, only the first 25 cycles were evaluated. Simulations with *hca* were not done.

7.2.2.1. Influence of the foundation stiffness

The substructure consists of a hollow reinforced concrete frame which will be filled with ballast after installation on site. Instead of modelling the structure that detailed, a smeared elastic stiffness (bulk modulus) is assumed in the simulations.

Figure 7.8 shows a numerical comparison of the computed settlement of the cross-foundation without skirts for two different stiffnesses. The upper two figures show the settlement of two feet at rest immediately after installation, the lower two figures the settlement at maximum load. The simulations were done for the afore mentioned offshore site Borkum West, for which a different hypoplastic parameter set was employed. One can see that the average settlement after installation are independent of the foundation stiffness. This is reasonable, since the settlement are governed by the weight of the structure. However, the stress distribution in the interface is different since the plates of the softer foundation are more flexible. Contrary to that, a significant

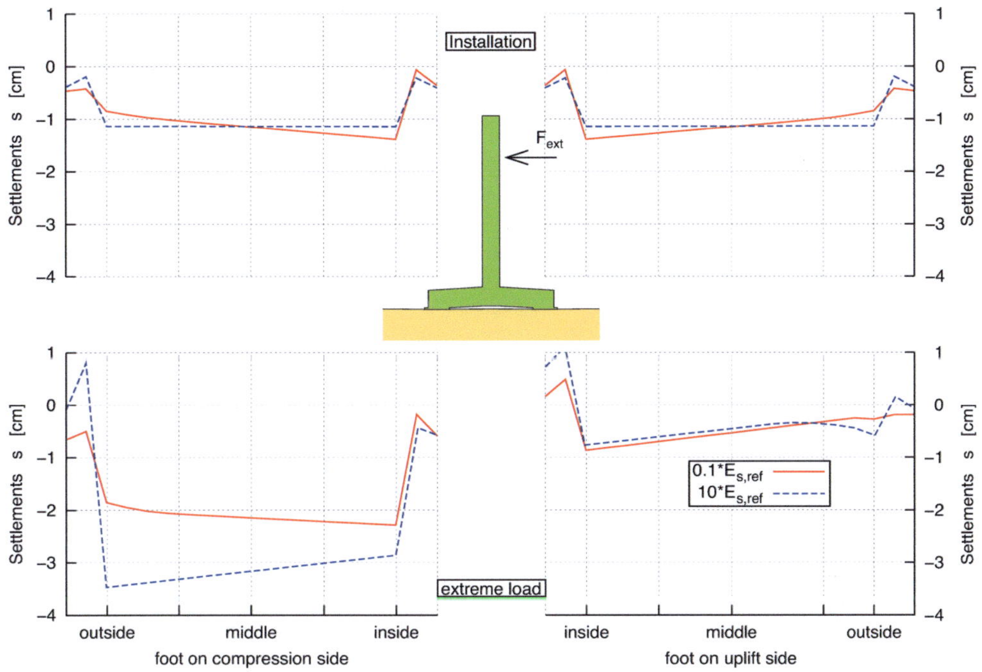

Figure 7.8.: Predicted settlements of the left and right foot after installation and at maximum load

influence on the settlement behaviour can be observed during extreme loading. This can be explained by means of local stress peaks on the compression side of the stiffer model which yields to larger deformations of the soil beneath the foot; lower left figure. In contrast to that, the flexible plate of the softer model allows a redistribution of the stresses in the interface, which yields to a small and uniform settlement.

The computed back-rotation, i.e. the average accumulated rotation after 25 cycles, of the cross-foundation for different stiffnesses, is shown in Figure 7.9; the corresponding settlement in Figure 7.10. The values are normalised according to Equation (5.2). The reference model has no skirts, i.e. $h_i = 0$ m, and has a stiffness of $E_{s,ref} = 2.5$ kN/mm^2. One can see, that both, normalised rotation and normalised settlement, are almost independent of the foundation stiffness. A detailed inspection in conjunction with Figure 7.8 reveals, that the amount of maximum $\hat{\varphi}_{max}$, residual $\hat{\varphi}_0$ and back-rotation $\Delta\hat{\varphi}^{av}$ increases with increasing foundation stiffness. Hence, the quotient $\Delta\hat{\varphi}^{av}/\hat{\varphi}_0$ is almost constant. This observation is, however, not valid for the average settlement. A possible explanation could be the partially impeded ability of the foundation to spread, i.e. the relative horizontal mobility of the feet. This explanation will be supported in the next section.

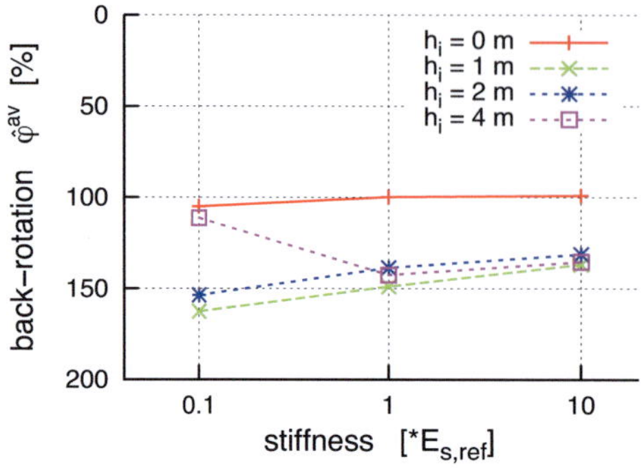

Figure 7.9.: Rotational behaviour of the cross foundation for different stiffnesses

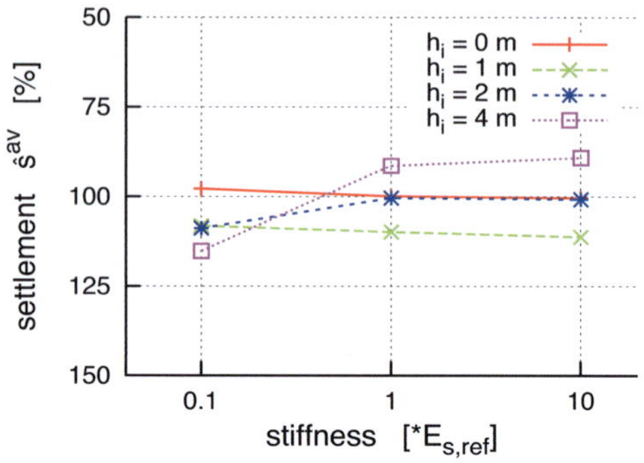

Figure 7.10.: Settlement behaviour of the cross foundation for different stiffnesses

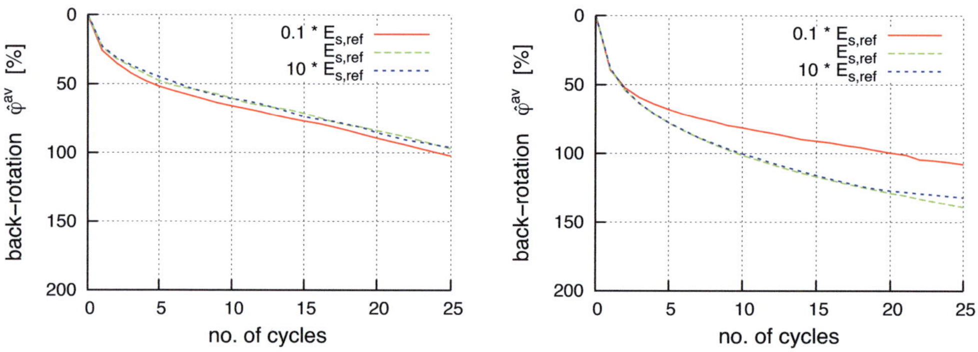

(a) Without skirts ($H_i = 0$ m)

(b) With skirts ($H_i = 4$ m)

Figure 7.11.: Comparison of the rotational behaviour for different substructure stiffnesses

The evolutions of the normalised back-rotation of the cross-foundation without and with 4 m long skirts, is presented in Figure 7.11. The observed independence of the back-rotation on the foundation stiffness, stated from Figure 7.10, is confirmed. Only the simulation of the soft foundation ($E_s = 0.1 \cdot E_{s,\text{ref}}$) and 4 m long skirts deviates from the other results. A reasonable mechanical explanation can hardly be found. It is most likely a numerical issue of the interface modelling between skirts and soil. Both bodies have exactly the same discretisation and a similar incremental stiffness. Under these conditions, the computation of both contact stresses and displacement is numerically difficult and may cause this deviating behaviour.

7.2.2.2. Influence of the skirt length

Figure 7.12 shows the absolute and normalised rotation of the cross-foundation for different skirt length. As already seen from Figure 7.9, the skirt length has only a small influence on the back-rotational behaviour, see Figure 7.13. The stabilisation is intensified by the presence of skirts, independent of the length. That means, that skirts reduce the residual rotation during extreme loading, Figure 7.12, and accelerate in the subsequent cyclic loading step the amount of accumulated back-rotation. The settlement is neither influenced by the presence of skirts nor by their length, Figure 7.14). Actually, the impression arises, that the accumulated average settlement even slightly decrease with increasing skirt length. Again, a deviating result can be observed for the soft foundation with 4 m long skirts.

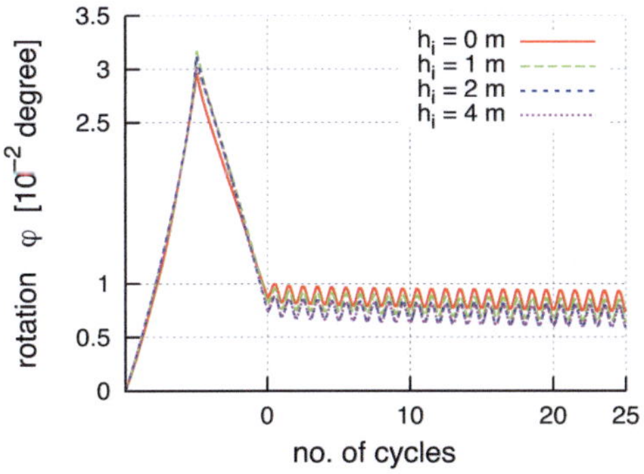

(a) Absolute rotation

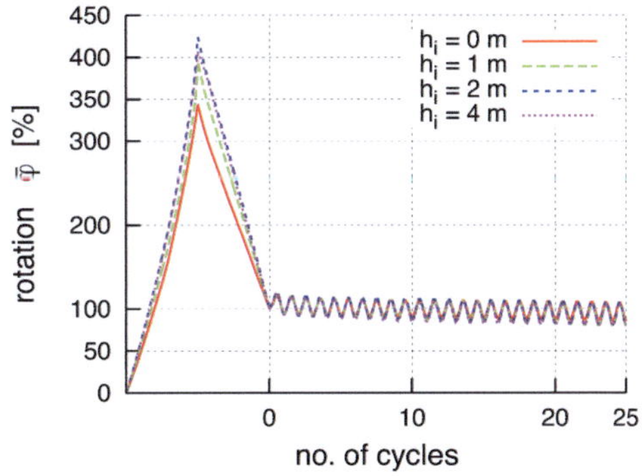

(b) Correlated rotation

Figure 7.12.: Computed rotation of the cross foundation incorporating different skirt length

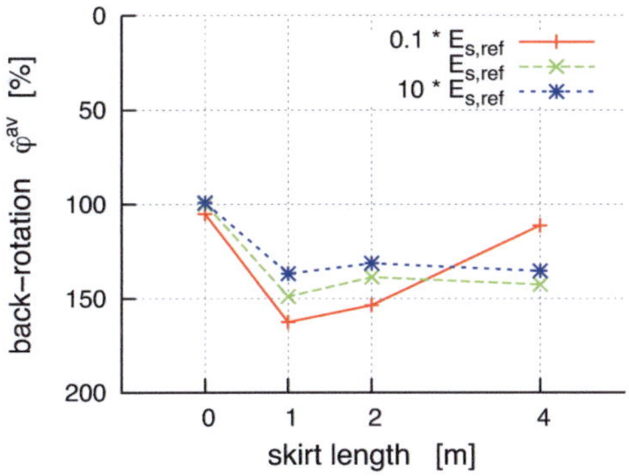

Figure 7.13.: Rotational behaviour of the cross foundation for different skirt length

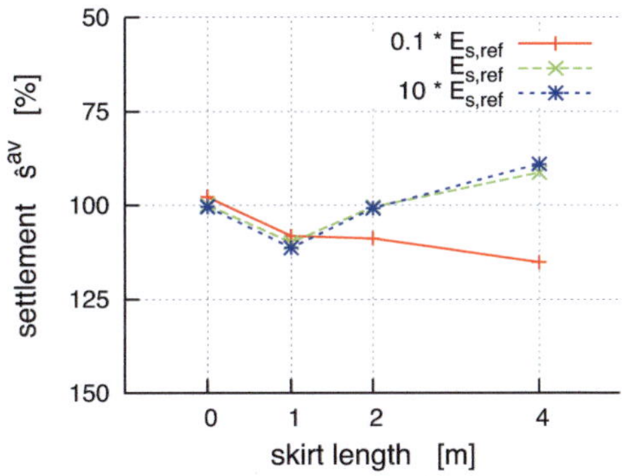

Figure 7.14.: Settlement behaviour of the cross foundation for different skirt length

7.3. Conclusion and Outlook

The presented new foundation structure represents a reliable solution for the foundation of offshore wind turbines. The important ability of a (shallow) foundation to stabilise itself during cyclic loading after a previous tilt due to an extreme loading, i.e. a severe storm or a large wave, is even slightly intensified by the presence of sufficiently deep skirts. The foundation stiffness has no influence on the overall stabilisation, but influence the amount of maximum rotation.

Besides the analysed behaviour, many other issues are of interest and could be the object of further studies. The positive effect of the skirts on the bearing capacity and the serviceability of the cross foundation emerges firstly, if the influence of transient pore water flow is taken into account. The buckets can mobilise significant amounts of suction on the uplift side which acts as a restoring force.

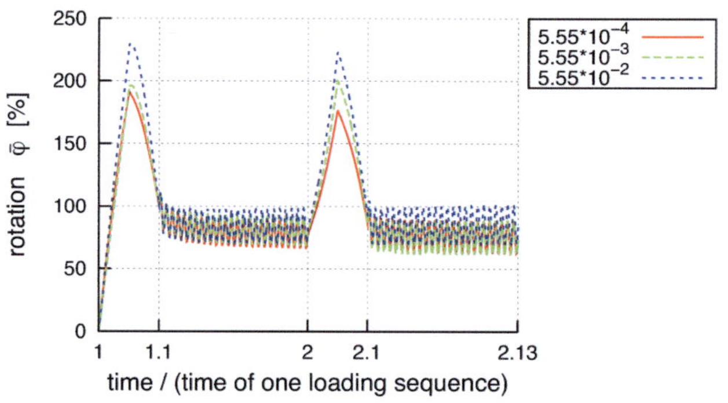

Figure 7.15.: Rotational behaviour of the cross foundation for loading conditions in accordance to the model tests on saturated soil and different permeabilities.

Figure 7.15 shows the influence of the permeability of a saturated soil on the stabilisation behaviour during extreme and subsequent cyclic loading of the cross-foundation without skirts. The simulated loading conditions correspond to the model tests; see Chapter 7.2.1. One can see, that the maximum rotation $\bar{\varphi}_{\max}$ decreases and the accumulated back-rotation $\Delta\bar{\varphi}^{\mathrm{av}}$ increases with decreasing permeability. The reason for that is the accumulated suction on the uplift side. It prevents the foundation from a large rotation during the extreme loading and accelerates the back-rotation. Other simulations with foundations incorporating skirts, not presented here, showed, that this behaviour is further intensified, since the amount of suction increases due to longer drainage paths. On the other hand, the positive effect of accumulated pore pressure vanishes for completely undrained soils. Under these conditions, the soil on the compression side looses its strength and the foundation tilts excessively during extreme loading.

8. Summary and outlook

8.1. Summary

It has been shown, that the entire spectrum of cyclic loads should be considered in the design of shallow foundations for offshore wind turbines. In particular, small load reversals during regular operation subsequent to an intensive load event can have a stabilising effect on the foundation. A critical design issue of support structures is the allowable rotation. Established calculation procedures are not yet capable to provide a reliable prediction. Uneconomical and insecure foundations are most likely, if only a few intensive load events are considered in the design.

The observed and systematically analysed back-rotational behaviour of shallow foundation during small cyclic loads, is both, a proof of serviceability and stability. A soil-mechanical explanation model is proposed which describes by means of results of laboratory element tests a mechanism causing the back-rotation. The model does not contradict the established *Palmgren-Miner* rule, which is employed in many disciplines dealing with millions of load reversals.

The mechanism of back-rotation, the proposed explanation and the employed constitutive models are additionally substantiated by an advanced physical soil model. By means of the total free energy and a gradient field of the so-called *granular temperature*, the stabilisation behaviour is physically supported. The back-rotation is not only a necessary, but actually a sufficient condition for the stability of a structure with a shallow foundation. On the other hand, an increase of accumulated rotation during small cyclic loads after an intensive load event means a lost of the stability of the foundation. From an energetic point of view, the total free energy trespasses a concave boundary, which will result in a collapse of the structure. A collapse is likely e.g. for a shallow foundation with a high seated centre of gravity. It could, however, not shown in frame of this work, since unstable boundary value problems cannot be captured by the employed numerical procedure.

Based on the results of the presented model tests and the numerical parameter study, an optimised geometry has been proposed. It allows the application of shallow foundations, which have several advantages compared to other foundation types, in deeper waters.

8.2. Outlook

The presented novel behaviour of shallow foundations under the considered loading conditions, provides a basis for many further studies, viz.

- the amount of back-rotation should be quantified in order to provide a reliable calculation procedure for support structures for offshore wind turbines;
- in particular, the influence of accumulated pore pressure should be considered in more detail. First results of coupled simulations, done in frame of this work, indicate that the back-rotational behaviour is even intensified, if the drainage conditions confine the accumulation of pore water pressure;
- complement model tests on shallow foundations subjected to combined cyclic loading in different directions could provide further inside into the mechanical behaviour, and would serve for a validation of the 2d simulations;
- in order to provide a reliable calculation procedure, in-situ measurements on installed offshore wind turbine structures are absolutely necessary;
- other foundation types should also be analysed with respect to the ability to back-rotate. Small-scale model tests on monopiles at the institute of soil- and rock-mechanics (IBF) in Karlsruhe reveal already a similar behaviour. Further studies are currently under way;
- an important tool in the design are numerical simulations. Thereto, advanced constitutive models are necessary. Thus, both, the employed high cycles accumulation model and the novel granular solid hydrodynamic model should be further developed. They have to be validated by large-scale model tests and in-situ measurements.

A. Design of the model foundation

A.1. Background

An octahedral shallow foundation was used in the model tests described in Chapter 4. A similar shaped foundation has already been employed to found offshore wind turbines in the Danish offshore wind park NYSTED [117] – but with with a hexagonal shape. The required (outer) diameter of the octahedral shallow foundation has been determined simplified on a circular plate in prototype dimensions with equivalent footprint. The design bases on two criteria,

- the safety against bearing failure;
- the avoidance of a opening gap by means of the limitation of the eccentricity.

Both criteria were calculated in accordance to German standards. The governing loads in prototype dimensions and the soil parameters are listed in Table A.1. They were determined for a cross-shaped shallow foundations at the planed German offshore wind park *Borkum West*. Further loads are listed in Figure C.1.

Table A.1.: Design loads and soil properties used to determine the prototype dimensions of the model foundation

ULS loads	
momentum at sea bed level $M_{y,k}$	542043.65 kNm
horizontal load at sea bed level $H_{x,k}$	9681.85 kN
vertical load (turbine and shaft) V_k	7500 kN
vertical load (foundation) G_k	(depends on the diameter and thickness)

Soil parameters	
submerged weight γ'	10 kN/m^3
friction angle φ_k	32°
cohesion c_k	0 kPa

The largest expected load, denoted as *ultimate limit state* (ULS) load, based on the load table shown in Figure C.1 has been adopted in the design. This is not necessarily the governing load case. Other combinations could be more critical, which depends mainly on the loading frequency, the foundation geometry and the drainage condition of the soil; i.e. the drainage path and permeability. If a significant pore pressure has been accumulated under the foundation due to previous (cyclic) loading, the conditions shown in Table A.1 may overestimate the bearing capacity. But in case of a plate-foundation, as discussed in the following, we will see, that not the bearing capacity but the avoidance of an open gap determines the required diameter. Hence it is reasonable to use the loads listed in Table A.1.

The loads given in Figure C.1 apply to the upper edge of an 8 m thick foundation plate. Hence, they have to be adjusted for the design with respect to the ground level, viz.

$$\begin{aligned} M_{\text{wave}} &= \left(\frac{283551.2}{8084.2}+8\right)\cdot 8084.2 \text{ kN/m} \\ &= 348224.8 \text{ kN/m} \\ M_{\text{wind}} &= \left(\frac{181037.65}{1597.65}+8\right)\cdot 1597.65 \text{ kN/m} \\ &= 193818.85 \text{ kN/m} \\ \Rightarrow \quad M_{\text{GOK}} &= M_{\text{wave}}+M_{\text{wind}} \\ &= \underline{\underline{542043.65 \text{ kN/m}}} \end{aligned}$$

In order to determine the required diameter of the model foundation, a thickness of the plate has to be assumed. The foundation will be most likely a reinforced concrete frame filled with ballast, such as gravel or stones. Based on the foundation used to determine the loads shown in Figure C.1, an average thickness of $\bar{h} = 7$ m is assumed.

A.2. Determination of the diameter of the foundation

A.2.1. Bearing capacity

Determination of the reduced diameter b':

$$b' = b - 2\cdot e_x = b - 2\cdot\frac{M_k}{V_k+G_k}$$

determination of the bearing capacity coefficients N_{d0} und N_{b0}:

$$
\begin{aligned}
N_{d0} &= \tan^2\left(45+\frac{\varphi}{2}\right) \\
N_{b0} &= (N_{d0}-1)\cdot\tan(\varphi) \\
N_{b0} &= \left[\tan^2\left(45+\frac{\varphi}{2}\right)-1\right]\cdot\tan(\varphi)
\end{aligned}
$$

determination of the shape coefficient ν_b:

$$
\nu_b = 1+\sin(\varphi)
$$

determination of the load angle δ:

$$
\begin{aligned}
\tan(\delta) &= \frac{H_k}{N_k} \\
\tan(\delta) &= \frac{H_k}{V_k+G_k}
\end{aligned}
$$

determination of the load angle coefficient i_b:

$$
\begin{aligned}
\omega &= 90^\circ \quad \Rightarrow \quad m = m_b \\
m &= \left(2+\frac{b'}{a'}\right)\cdot\left(1+\frac{b'}{a'}\right)^{-1} \\
&\approx \left(2+\frac{b'}{b}\right)\cdot\left(1+\frac{b'}{b}\right)^{-1} \\
i_b &= [1-\tan(\delta)]^{m+1} \\
&\approx [1-\tan(\delta)]^{\left(\frac{2+\frac{b'}{b}}{1+\frac{b'}{b}}+1\right)}
\end{aligned}
$$

determination of the resistance factors:

$$
\begin{aligned}
N_b &= N_{b0}\cdot\nu_d\cdot i_d \\
R_{n,k,1} &= b'\cdot a'\cdot\left(\gamma'\cdot b'\cdot N_b\right) \\
&\approx b'^2\cdot b\cdot\gamma'\cdot N_{b0}\cdot\nu_d\cdot i_d
\end{aligned} \tag{A.1}
$$

an the safety against bearing failure:

$$
\begin{aligned}
N_d &\leq R_d \\
N_k \cdot \gamma_{\mathrm{G}} &\leq \frac{R_{n,k}}{\gamma_{\mathrm{Gr}}} \\
\Rightarrow \quad R_{n,k,2} &= N_k \cdot \gamma_{\mathrm{G}} \cdot \gamma_{\mathrm{Gr}} \\
&= (V_k + G_k) \cdot \gamma_{\mathrm{G}} \cdot \gamma_{\mathrm{Gr}} \\
&= (V_k + G_k) \cdot 1.35 \cdot 1.4 \\
&= 1.89 \cdot (V_k + G_k)
\end{aligned} \tag{A.2}
$$

The vertical static load from the foundation can be determined viz.:

$$
G_k = \pi \cdot \left(\frac{b}{2}\right)^2 \cdot h \cdot \gamma_c'
$$

with γ_c' being the average specific volume weight. Equation (A.1) and Equation (A.2) combined:

$$
\begin{aligned}
R_{n,k} &= R_{n,k} \\
1.89 \cdot (V_k + G_k) &= A' \cdot (\gamma' \cdot b' \cdot N_b) \\
1.89 \cdot \left[V_k + \left(\pi \cdot \left[\frac{b}{2}\right]^2 \cdot h \cdot \gamma_c'\right)\right] &= \left(b - 2 \cdot \frac{M_k}{V_k + \left[\pi \cdot \left(\frac{b}{2}\right)^2 \cdot h \cdot \gamma_c'\right]}\right) \cdot A' \cdot \gamma' \cdot N_b
\end{aligned} \tag{A.3}
$$

with

$$
\begin{aligned}
N_b &= N_{b0} \cdot \nu_d \cdot i_d \\
&= \left(\left[\tan^2\left(45 + \frac{\varphi}{2}\right)\right] - 1\right) \cdot \tan(\varphi) \cdot [1 + \sin(\varphi)] \\
&\quad \cdot \left[1 - \frac{H_k}{V_k + \left[\pi \cdot \left(\frac{b}{2}\right)^2 \cdot h \cdot \gamma_c'\right]}\right]^{\left(\frac{2 + \frac{\left[b - 2 \cdot \frac{M_k}{V_k + \left(\pi \cdot \left[\frac{b}{2}\right]^2 \cdot h \cdot \gamma_c'\right)}\right]}{b}}{1 + \frac{\left[b - 2 \cdot \frac{M_k}{V_k + \left(\pi \cdot \left[\frac{b}{2}\right]^2 \cdot h \cdot \gamma_c'\right)}\right]}{b}} + 1\right)}
\end{aligned}
$$

A.2.2. Limiting of the eccentricity

The eccentricity has to be limited according to BOROWICKA [26] for circular plates to $e_x \leq b/6$.

$$
\begin{aligned}
b &= 6 \cdot e_x \\
&= 6 \cdot \frac{M_k}{V_k + G_k} \\
&= 6 \cdot \frac{M_k}{V_k + \pi \cdot \left(\frac{b}{2}\right)^2 \cdot h \cdot \gamma_c'}
\end{aligned} \tag{A.4}
$$

The reduced effective footprint can be calculated viz.

$$
\begin{aligned}
A' &= 2 \cdot F \\
&= 2 \cdot \frac{(b/2)^2}{2} \cdot \left(\frac{2 \cdot \alpha}{180} - \sin(\alpha)\right) \\
&= 2 \cdot \frac{(b/2)^2}{2} \cdot \left(\frac{\pi \cdot \left(2 \cdot \arccos\left(\frac{1}{6}\right)\right)}{180} - \sin\left(2 \cdot \arccos\left(\frac{1}{6}\right)\right)\right) \\
&= (b/2)^2 \cdot \left(\frac{\pi \cdot \arccos\left(\frac{1}{6}\right)}{90} - \sin\left(2 \cdot \arccos\left(\frac{1}{6}\right)\right)\right) \\
&\approx b^2 \cdot 0.458
\end{aligned}
$$

A.2.3. Prototype and model dimensions

The required diameter b is the maximum value of Equation (A.3) and Equation (A.4). With increasing water depth and turbine capacity, the overturning moment at ground level will increase too. Hence Equation (A.4) will yield larger to larger values than Equation (A.3). An optimised foundation geometry, however, should deliver similar values for both equations. Thus, thin plates, compared to the diameter of a foundation, are uneconomical. In order to improve the geometry, one could use thicker plates which will lead to larger vertical stresses in the soil-structure-interface and hence smaller diameters. But thicker foundation require taller structures with higher centre of gravity. And this geometry is not favourable for a stabilisation (self-healing) as discussed in Chapter 6.

The dimensions of the octahedral shallow foundation can be calculated viz.

$$
\begin{aligned}
\pi \cdot r_k^2 &\overset{!}{=} 2 \cdot \sqrt{2} \cdot r_u^2 \\
\Rightarrow \quad r_u &= r_k \cdot \sqrt{\frac{\pi}{2 \cdot \sqrt{2}}} \\
s &= r_k \cdot \sqrt{\frac{\pi}{2 \cdot (\sqrt{2} + 1)}}
\end{aligned}
$$

with r_u denoting the outer radius and s denoting the secant length.

The final dimensions of the model foundation depends on the scaling factor. In this work a factor of 1:125 was chosen.

B. Detailed views of the model test rig

Figure B.1.: Control and measurement equipment: pneumatic unit at the right side, PC and data-locker on the left site.

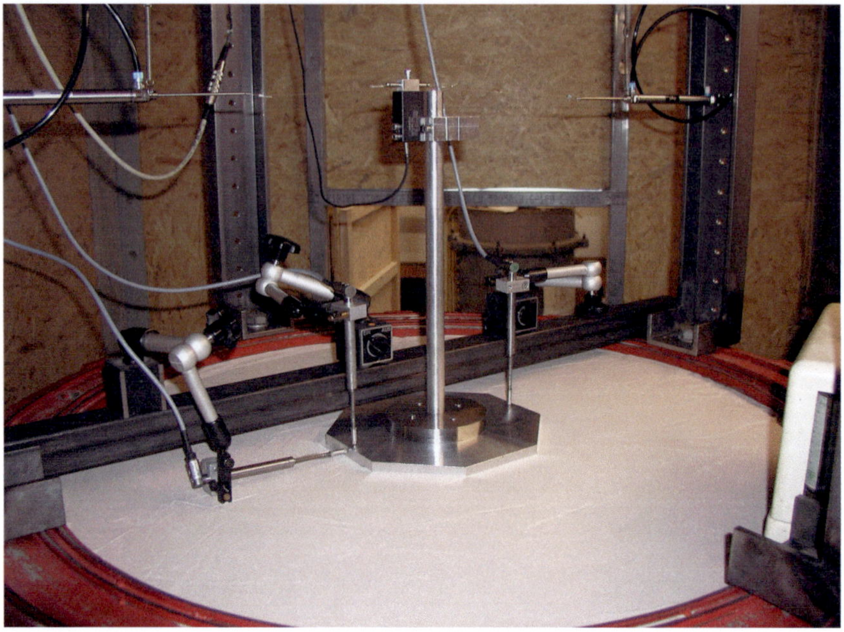

(a) View on model foundation and instrumentation

(b) View on the loading frame on top of the former triaxial cell

Figure B.2.: View of model foundation with shaft and load cell, 2 pneumatic cylinders, 2 vertical and 1 horizontal displacement transducers, loading frame and tentative measuring beam.

C. Load table of a representative Offshore Wind Park

Figure C.1 shows a list of characteristic loads determined for a shallow foundation at the planed German offshore wind park *Borkum West*. The upper part lists the loads which should be used in order to determine the so-called *fatigue limit state* (FLS); the lower part lists the load for the *ultimate limit state* (ULS). The FLS loads are sorted in packages with equal amplitude and average value. The number of repetitions decrease with increasing amplitude.

The table is divided into a left and right table. On the left side are listed the loads from wave and current, and on the right side the static vertical and alternating horizontal loads from the turbine (wind).

load combinations for FLS	wave loads: wave height $H_{s,wave}$ [m]	wave loads: horizontal load $F_{h,wave}$ [kN]	wave loads: moment load $M_{b,wave}$ [kN]	number [N]	turbine loads (at upper edege of foundation): horizontal load $F_{h,turbine}$ [kN]	turbine loads (at upper edege of foundation): vertical load $F_{v,turbine}$ [kN]	turbine loads (at upper edege of foundation): moment load $M_{b,turbine}$ [kNm]	turbine loads (at upper edege of foundation): torsion moment $M_{t,turbine}$ [kNm]	number [N]
LK 1	1,50	1118,57	38697,52	3,839E+07	852,12	7500,00	100797,04	3000,00	2,311E+05
LK 2	1,60	1167,83	39100,73	2,220E+07	852,12	7500,00	100797,04	3000,00	1,547E+05
LK 3	2,00	1586,43	52796,27	2,027E+07	852,12	7500,00	100797,04	3000,00	1,547E+05
LK 4	2,40	1748,80	56929,50	1,664E+07	985,00	7500,00	109541,00	7000,00	1,382E+05
LK 5	2,80	1750,60	55393,40	1,247E+07	985,00	7500,00	109541,00	7000,00	1,116E+05
LK 6	3,20	1787,00	54902,90	8,534E+06	985,00	7500,00	109541,00	7000,00	8,207E+04
LK 7	3,60	1853,50	55256,60	5,377E+06	985,00	7500,00	109541,00	7000,00	5,521E+04
LK 8	4,00	1926,80	55712,10	3,115E+06	985,00	7500,00	109541,00	7000,00	3,405E+04
LK 9	4,40	1969,30	56212,80	1,663E+06	985,00	7500,00	109541,00	7000,00	1,929E+04
LK 10	4,80	2061,00	57172,60	8,199E+05	985,00	7500,00	109541,00	7000,00	1,004E+04
LK 11	5,00	2137,40	57708,40	3,724E+05	985,00	7500,00	109541,00	7000,00	4,810E+03
LK 12	5,60	2206,30	59506,90	1,562E+05	985,00	7500,00	109541,00	7000,00	2,119E+03
LK 13	6,00	2295,60	60501,30	1,615E+05	985,00	7500,00	109541,00	7000,00	2,299E+03
sum				1,302E+08					1,000E+06

load combinations for ULS	wave height $H_{s,wave}$ [m]	horizontal load $F_{h,wave}$ [kN]	moment load $M_{b,wave}$ [kN]	number [N]	horizontal load $F_{h,turbine}$ [kN]	vertical load $F_{v,turbine}$ [kN]	moment load $M_{b,turbine}$ [kNm]	torsion moment $M_{t,turbine}$ [kNm]	number [N]
LK 1	18,90	8084,20	283551,20	1,000E+00	1597,65	7500,00	181037,65	14025,00	1,000E+00

Figure C.1.: Characteristic load table determined for a shallow foundation at the planed offshore wind park *Borkum West*

D. Convergence study

In order to study the influence of the element type, discretisation, i.e. mainly the element size, and the increment size, comparative simulations have been done. They are presented in the following figures. The comparison is made by means of the amount of relative back-rotation. The rotation is referred to the rotation of the model which has been used in the simulations shown in Chapter 5. The dimensions of the model and the mesh geometry was equal in all simulations. Similar comparisons have been done for the finite element model of the cross-foundation.

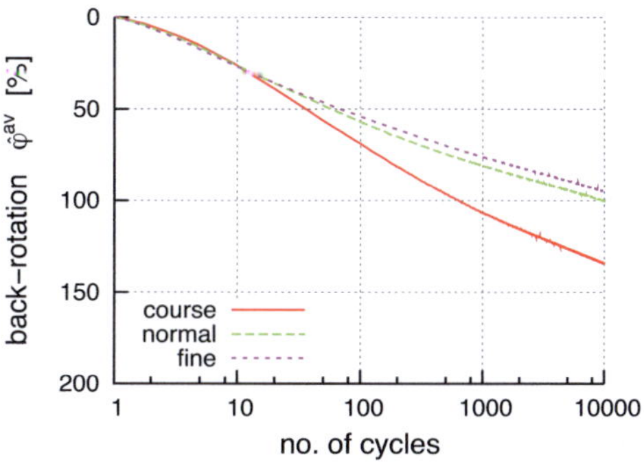

Figure D.1.: Comparison of different element size. The number of elements is 4 times larger in the *fine* model and 4 times less in the *course* model, both referred to the *normal* model.

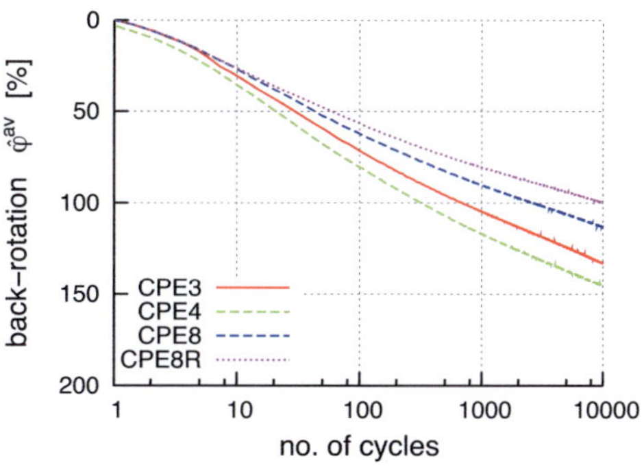

Figure D.2.: Comparison of different element types (names of ABAQUS elements).

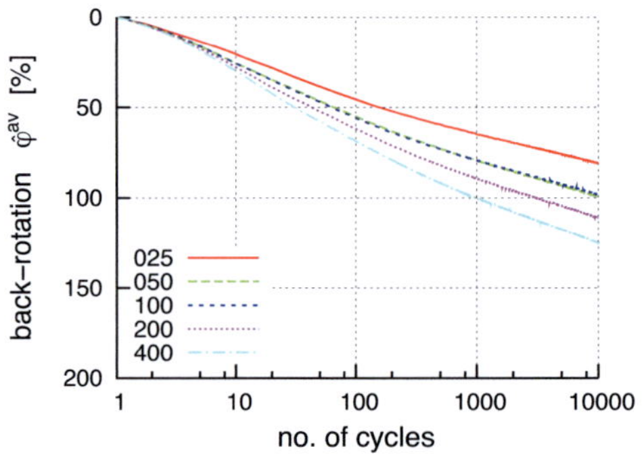

Figure D.3.: Variation of number of load cycles computed in one increment in the high cycle accumulation model.

E. Bibliography

[1] P. Aas and K. Andersen. Skirted foundations for offshore structures. In *Offshore South East Asia - 9th Conference and Exhibition*, pages 305–312, 1992.

[2] M. Allard. NGI-DG Centrifuge tests of Ekofisk type oil storage tank on very dense sand bed. Technical Report SE-51043/32 and CO-344070/36, Delft Geotechnics, 1993.

[3] M. Allard, K. Andersen, and J. Hermstad. Centrifuge model tests of a gravity platform on very dense sand; Part I: Testing technique and results. In *BOSS'94*, volume 1, pages 231–254, 1994.

[4] N. Allotey and M. El Naggar. A consistent soil fatigue framework based on the number of equivalent cycles. *Geotechnical and Geological Engineering*, 26(1):65–77, Februar 2008. doi: 10.1007/s10706-007-9147-2.

[5] Alpha Ventus. Alpha Ventus, 2009. URL `http://www.alpha-ventus.de/`.

[6] Altair Engineering. Hyperworks, 2009. URL `http://www.altair.com/`.

[7] American Petroleum Institute. API RP 2A-LRFD - Recommended practice for planning, designing and constructing fixed offshore platforms - Load and resistance factor design, July 1993.

[8] American Petroleum Institute. API RP 2A-WSD - Recommended practice for planning, designing and constructing fixed offshore platforms - Working stress design, December 2000.

[9] American Society for Testing and Materials. ASTM D4254 - Standard test methods for minimum index density and unit weight of soils and calculation of relative density, 2006.

[10] American Society for Testing and Materials (ASTM). ASTM D4253 - Standard test methods for maximum index density and unit weight of soils using a vibratory table, 2006.

[11] K. Andersen. Foundation design of offshore gravity structures. In M. O'Reilly and S. Brown, editors, *Cyclic Loading of Soils - From Theory to Design*, pages 122–173. Blackie and Son Ltd., 1991.

[12] K. Andersen. Cyclic clay data for foundation design of structures subjected to wave loading. In T. Triantafyllidis, editor, *International Conference on Cyclic Behaviour of Soils and Liquefaction Phenomena*, pages 371–387, Bochum, 2004.

[13] K. Andersen. Bearing capacity under cyclic loading - offshore, along the coast, and on land. The 21st Bjerrum Lecture presented in Oslo, 23 November 2007. *Géotechnique*, 46(5):513–535, 2009. doi: 10.1139/T09-003.

[14] K. Andersen and T. Berre. Behaviour of a dense sand under monotonic and cyclic loading. In *12th European Conference on Soil Mechanics and Foundation Engineering*, volume 2, pages 667–676, Amsterdam, 1999.

[15] K. Andersen and H. Jostad. Shear strength along inside of suction anchor skirt wall in clay. In *OTC - Offshore Technology Conference*, 2004.

[16] K. Andersen, M. Allard, and J. Hermstad. Centrifuge model tests of a gravity platform on very dense Part II: Interpretation. In *BOSS'94*, volume 1, pages 255–282, 1994.

[17] K. Andersen, J. Murff, M. Randolph, E. Clukey, C. Erbrich, H. Jostad, B. Hansen, C. Aubeny, P. Sharma, and C. Supachawarote. Suction anchors for deepwater applications. In S. Gourvenec and J. Cassidy, editors, *ISOFOG - International Symposium on Frontiers in Offshore Geotechnics*, 2005.

[18] K. Andersen, H. Jostad, and R. Dyvik. Penetration resistance of offshore skirted foundations and anchors in dense sand. *Journal of Geotechnical and Geoenvironmental Engineering*, 134(1):106–116, January 2008.

[19] L. Bardon. Time dependent deformations of normally consolidated clays and peats. *Journal of the Soil Mechanics and Foundations Division*, 95(1):1–31, January 1969.

[20] E. Bauer. *Zum mechanischen Verhalten granularer Stoffe unter vorwiegend ödometrischer Beanspruchung*. PhD thesis, Publication of the Institute of Soil and Rock Mechanics in Karlsruhe, 1992.

[21] E. Bauer. Calibration of a comprehensive hypoplastic model for granular materials. *Soils and Foundations*, 36(1):13–26, 1996. URL `http://ci.nii.ac.jp/naid/110003945985/en/`.

[22] T. Berre and L. Bjerrum. Shear strength of normally consolidated clays. In *8th International Conference on Soil Mechanics and Foundation Engineering*, pages 39–49, Moscow, 1973.

[23] A. Bishop. The strength of soils as engineering materials - 6th Rankine Lecture. *Géotechnique*, 16(2):91–128, 1966.

[24] L. Bjerrum. Geotechnical problems involved in foundations of structures in the North Sea. *Géotechnique*, 23(3):319–358, 1973.

[25] BMU. Erneuerbare-Energien-Gesetz (EEG 2004), 2004. URL http://www.bmu.de/erneuerbare/energien/doc/5982.php.

[26] H. Borowicka. Über ausmittig belastete, starre Platten auf elastisch-isotropem Untergrund. *Archive of Applied Mechanics (Ingenieur Archiv)*, 14(1):1–8, 1943. URL http://dx.doi.org/10.1007/BF02084318.

[27] G. Bouckovalas, R. Whitman, and W. Marr. Permanent displacement of sand with cyclic loading. *Journal of Geotechnical Engineering*, 110(11):1606–1623, 1984. doi: 10.1061/(ASCE)0733-9410(1984)110:11(1606). URL http://link.aip.org/link/?QGE/110/1606/1.

[28] G. Bouckovalas, W. Marr, and J. Christian. Analyzing permanent drift Due to cyclic loads. *Journal of Geotechnical Engineering*, 112(6):579–593, 1986. doi: 10.1061/(ASCE)0733-9410(1986)112:6(579). URL http://link.aip.org/link/?QGE/112/579/1.

[29] BSH - Bundesamt für Seeschifffahrt und Hydrographie. Konstruktive Ausführung von Offshore-Windenergieanlagen. download, 2007. URL http://www.bsh.de/de/Produkte/Buecher/Standards_Windenergie/.

[30] J. Burland, M. Jamiolkowski, and C. Viggiani. Stabilising the leaning tower of Pisa. *Bulletin of Engineering Geology and the Environment*, 57(1):91–99, 1998.

[31] R. Butterfield. A natural compression law for soils. *Géotechnique*, 29, 1979.

[32] A. Bye, C. Erbrich, B. Rognlien, and T. Tjelta. Geotechnical design of Bucket foundations. In *OTC - Offshore Technology Conference*, 1995.

[33] B. Byrne and G. Houlsby. Drained behaviour of suction caisson foundations on very dense sand. In *OTC - Offshore Technology Conference*, 1999.

[34] B. Byrne and G. Houlsby. Experimental investigations of the response of suction caissons to transient combined loading. *Journal of Geotechnical and Geoenvironmental Engineering*, 130(3):240–253, March 2004.

[35] A. Casagrande. Liquefaction and cyclic deformation of sands: a critical review. In *Proceedings of the Pan-American Conference on Soil Mechanics and Foundation Engineering*, volume 5, pages 79–133, Buenos Aires, 1975.

[36] G. Castro. Liquefaction and cyclic mobility of saturated sands. *Journal of the Geotechnical Engineering Devision*, 101(6):551–569, 1975.

[37] G. Castro and S. Poulos. Factors affecting liquefaction and cyclic mobility. *Journal of the Geotechnical Engineering Devision*, 103(6):501–506, 1977.

[38] R. Chambon, J. Desrues, W. Hammad, and R. Charlier. CLoE, a new rate-type constitutive model for geomaterials theoretical basis and implementation. *International Journal for Numerical and Analytical Methods in Geomechanics*, 18(4):253–278, July 1994. doi: 10.1002/nag.1610180404.

[39] C. Clausen, E. DiBiagio, J. Duncan, and K. Andersen. Observed behaviour of the Ekofisk oil storage tank foundation. In *OTC - Offshore Technology Conference*, volume 3, pages 399–413, Houston, 1975.

[40] G. Clauss. The taming of the shrew - Tailoring freak waves for seakeeping tests. *Journal of Ship Research*, 52(3):194–226, September 2008. URL `http://www.ingentaconnect.com/content/sname/jsr/2008/00000052/00000003/art00004`.

[41] R. Cudmani. *Statische, alternierende und dynamische Penetration in nichtbindige Böden*. PhD thesis, Publication of the Institute of Soil and Rock Mechanics in Karlsruhe, 2001.

[42] R. Cudmani and H. Sturm. An investigation of the tip resistance in granular and soft soils during static, alternating and dynamic penetration. In H. Gonin, A. Holeyman, and F. Rocher-Lacoste, editors, *TransVib 2006: International Symposium on vibratory pile driving and deep soil compaction*, pages 221–231, 2006.

[43] Y. Dafalias and M. Manzari. Simple plasticity sand model accounting for fabric change effect. *Journal of Engineering Mechanics*, 130(6):622–634, 2004.

[44] Dassault Systèmes. ABAQUS, 2009. URL `http://www.simulia.com/`.

[45] M. de Groot, M. D. Bolton, P. Foray, P. Meijers, A. Palmer, R. Sandven, A. Sawicki, and T. Teh. Physics of liquefaction phenomena around marine structures. *Journal of Waterway, Port, Coastal, and Ocean Engineering*, 132(4):227–243, 2006. doi: 10.1061/(ASCE)0733-950X(2006)132:4(227). URL `http://link.aip.org/link/?QWW/132/227/1`.

[46] Det Norske Veritas. Offshore Standard DNV-OS-J101 - Design of Offshore Wind Turbine Structures, October 2007.

[47] Deutsche Energie-Agentur (dena). Willkommen bei www.Offshore-Wind.de, 2009. URL `http://www.offshore-wind.de`.

[48] Deutsches Institut für Normung e. V. (DIN). DIN 18126 - Bestimmung der Dichte nichtbindiger Böden bei lockerster und dichtester Lagerung, November 1996.

[49] Deutsches Institut für Normung e. V. (DIN). DIN 19700-11 - Stauanlagen Teil 11: Talsperren, July 2004.

[50] Deutsches Institut für Normung e. V. (DIN). DIN 1054 - Baugrund - Sicherheitsnachweise im Erd- und Grundbau, January 2005.

[51] Deutsches Institut für Normung e. V. (DIN). Eurocode EC7: Entwurf, Berechnung und Bemessung in der Geotechnik Teil1: Allgemeine Regel, October 2005.

[52] DEWI - German Wind Energy Institute. DEWI - German Wind Energy Institute, 2009. URL `www.dewi.de`.

[53] H. Di Benedetto, F. Tatsuoka, and M. Ishihara. Time-dependent shear deformation characteristics of sand and their constitutive modelling. *Soils and Foundations*, 42(2):1–22, April 2002. URL `http://ci.nii.ac.jp/naid/110003945829/en/`.

[54] V. Diyaljee and G. Raymond. Repetitive load deformation of cohesionless soils. *Journal of Geotechnical Engineering Devision*, 108(10):1215–1229, October 1982.

[55] O. Eide and K. Andersen. Foundation engineering for gravity structures in the northern North Sea. In *International Conference on Case Histories in Geotechnical Engineering*, volume 4, pages 1627–1678, 1984.

[56] C. Erbrich and T. Tjelta. Installation of bucket foundations and suction caissons in sand - Geotechnical performance. In *OTC - Offshore Technology Conference*, 1999.

[57] Germanischer Lloyd Windenergie GmbH. Guideline for the certification of offshore wind turbines, 2004.

[58] G. Gottardi, L. Govoni, and R. Butterfield. Yield loci for shallow foundations by swipe testing. In S. Gourvenec and J. Cassidy, editors, *ISOFOG - International Symposium on Frontiers in Offshore Geotechnics*, pages 469–475, 2005.

[59] J. Grabe, J. Dührkop, and K.-P. Mahutka. Monopilegründungen von Offshore Windenergieanlagen - Zur Bildung von Porenwasserüberdrücken aus zyklischer Belastung. *Bauingenieur*, 9(6):418–423, 2004.

[60] P. Grammenoudis. *Mikripolare Plastizität.* PhD thesis, Fachbereich Mechanik der Technischen Universität Darmstadt, 2003.

[61] G. Gudehus. A visco-hypoplastic constitutive relation for soft soils. *Soils and Foundations*, 44(4):11–25, August 2004. URL `http://ci.nii.ac.jp/naid/110003892441/en/`.

[62] G. Gudehus. Seismo-hypoplasticity with a granular temperature. *Granular Matter*, 8(2): 93–102, May 2005.

[63] G. Gudehus. Vibro-viscosity and granular temperature of cylindrical grain skeletons - Theory. In *Powders and Grains*, volume 1, pages 1087–1090. A.A. Balkema, 2005.

[64] P. Haff. Grain flow as a fluid-mechanical phenomenon. *Journal of Fluid Mechanics*, 134 (1):401–430, 1983.

[65] B. Hansen, F. Nowacki, E. Skomedal, and J. Hermstad. Foundation design, Troll platform. In *6th International Conference on the Behaviour of Offshore Structures*, volume 2, pages 921–936, London, 1992.

[66] HBM. HBM measurements, 2009. URL `http://www.hbm.com/`.

[67] I. Herle. *Hypoplastizität und Granulometrie einfacher Korngerüste.* PhD thesis, Publication of the Institute of Soil and Rock Mechanics in Karlsruhe, 1997.

[68] I. Herle and G. Gudehus. Determination of parameters of a hypoplastic constitutive model from properties of grain assemblies. *Mechanics of Cohesive-Frictional Materials*, 4:461–486, 1999.

[69] A. Hettler. *Verschiebungen starrer und elastischer Gründungskörper in Sand bei monotoner und zyklischer Belastung*. PhD thesis, Publication of the Institute of Soil and Rock Mechanics in Karlsruhe, 1981.

[70] U. Holzlöhner. Settlements of shallow foundations on sand. *Soils and Foundations*, 24 (4):58–70, December 1984.

[71] U. Holzlöhner. Fundamentsetzungen unter Gebrauchslast. In F. Rackwitz, editor, *Entwicklungen in der Bodenmechanik, Bodendynamik und Geotechnik*, pages 17–32, 2006.

[72] Horns Rev Havmøllepark. Horns rev havmøllepark, 2009. URL `http://www.hornsrev.dk`.

[73] G. Houlsby and B. Byrne. Design procedures for installation of suction caissons in clay Design procedures for installation of suction caissons in clay and other materials. *Proceedings of the Institution of Civil Engineers - Geotechnical Engineering*, 158:75–82, April 2005.

[74] G. Houlsby and B. Byrne. Design procedures for installation of suction caissons in sand. *Proceedings of the Institution of Civil Engineers - Geotechnical Engineering*, 158:135–144, July 2005.

[75] G. Houlsby and A. Puzrin. *Principles of hyperplasticity: An approach to plasticity theory based on thermodynamic principles*. Springer, 2006.

[76] G. Houlsby, L. Ibsen, and B. Byrne. Suction caissons for wind turbines. In S. Gourvenec and J. Cassidy, editors, *ISOFOG - International Symposium on Frontiers in Offshore Geotechnics*, pages 75–93, 2005.

[77] G. Huber and H. Wienbroer. Vibro-viscosity and granular temperature of cylindrical grain skeletons - Experiments. In *Powders and Grains*, volume 2, pages 287–290. A.A. Balkema, 2005.

[78] M. Hvorslev. *Über die Festigkeitseigenschaften gestörter bindiger Böden*. PhD thesis, Ingeniorvidenskabelige Skrifter A, Nr. 45, 1937.

[79] M. Hyodo, H. Tamimizu, N. Yasufuku, and H. Murata. Undrained cyclic and monotonic triaxial behaviour of saturated loose sand. *Soils and Foundations*, 34(1):19–32, March 1994. URL `http://ci.nii.ac.jp/naid/110003960029/en/`.

[80] L. Ibsen. The stable state in cyclic triaxial testing on sand. *Soil Dynamics and Earthquake Engineering*, 13(1):63–72, 1994.

[81] L. Ibsen and R. Brincker. Design of a new foundation for offshore wind turbines. In *Proceedings of the 22nd International Modal Analysis Conference (IMAC)* , Detroit, Michigan, 2004.

[82] G. Imai and Y. Tang. A constitutive equation of one-dimensional consolidation derived from inter-connected tests. *Soils and Foundations*, 32(2):83–96, June 1992. URL `http://ci.nii.ac.jp/naid/110003959890/en/`.

[83] K. Ishihara, F. Tatsuoka, and S. Yasuda. Undrained deformation and liquefaction of sand under cyclic stresses. *Soils and Foundations*, 15(1):29–44, March 1975. URL `http://ci.nii.ac.jp/naid/110003959119/en/`.

[84] J. Jenkins and S. Savage. A theory for the rapid flow of identical, smooth, nearly elastic, spherical particles. *Journal of Fluid Mechanics*, 130(1):187–202, 1983.

[85] Y. Jiang and M. Liu. From elasticity to hypoplasticity: Dynamics of granular solids. *Physical Review Letters*, 99(10):105501 1–4, 2007.

[86] H. Jostad, K. Andersen, and T. Tjelta. Analyses of skirted foundations and anchors in sand subjected to cyclic loading. In *International Conference on the Behaviour of Offshore Structures*, Delft, 1997.

[87] C. Karcher. *Tagebaubedingte Deformationen Tagebaubedingte Deformationen im Lockergestein*. PhD thesis, Publication of the Institute of Soil and Rock Mechanics in Karlsruhe, 2002.

[88] D. Kolymbas. *Ein nichtlineares viskoplastisches Stoffgesetz für Böden*. PhD thesis, Publication of the Institute of Soil and Rock Mechanics in Karlsruhe, 1978.

[89] D. Kolymbas. *Geotechnik*. Springer, 2007.

[90] H. Koreck and P. Schwarz. Axial cyclic loaded piles. In *International Geotechnical Seminar on Deep Foundations on Bored and Auger Piles*, pages 395–399, Ghent, June 1988. Balkema.

[91] S. Lacasse and T. D'Orazio. Soil reaction stresses on gravity platforms. In *12th International Conference on Soil Mechanics and Foundation Engineering*, volume 1, pages 445–448, 1989.

[92] S. Lacasse and L. Robberstad. Foundation behaviour of the Frigg CDP1 Platform. In *BOSS 92, 6th International Conference on the Behaviour of Offshore Structures*, volume 1, pages 95–106. BPP Tech Services Ltd, July 1992.

[93] S. Lacasse, A. Goulois, L. Robberstad, E. Andersen, and P. Boisard. The foundation of the Frigg CDP1 gravity platform: A case study. In *OTC - Offshore Technology Conference*, volume 1, pages 125–131, 1991.

[94] T. Lambe, J. Boehmer, and W. Rosenbrand. Caisson tests at Neeltje Jans - Oosterschelde storm surge barrier project. Technical report, Rijkwaterstaat Deltadienst, 1977.

[95] J. Lanier, D. Caillerie, R. Chambon, G. Viggiani, P. Bésuelle, and J. Desrues. A general formulation of hypoplasticity. *International Journal for Numerical and Analytical Methods in Geomechanics*, 28(15):1461–1478, 2004. doi: 10.1002/nag.394.

[96] H. Leinenkugel. *Deformations- und Festigkeitsverhalten bindiger Erdstoffe. Experimentelle Ergebnisse und ihre physikalische Deutung.* PhD thesis, Publication of the Institute of Soil and Rock Mechanics in Karlsruhe, 1976.

[97] K. Mallwitz and U. Holzlöhner. Verfahren zur Ermittlung der Setzung von Fundamenten infolge zyklischer Belastung. *Bautechnik*, 73(3):175–186, March 1996.

[98] H. Matsuoka and T. Nakai. A new failure criterion for soils in three-dimensional stresses. In *IUTAM - Conference on Deformation and Failure of Granular Materials*, volume 1, pages 253–2263, Delft, 1982.

[99] M. Matsushita, F. Tatsuoka, J. Koseki, B. Cazacliu, H. di Benedetto, and S. Yasin. Time effects on the pre-peak deformation properties of sands. In *International Conference on Pre-Failure Deformation Characteristics of Geomaterials*, pages 681–689. Balkema, 1999.

[100] B. Mazurkiewicz and M. Topolnicki. Gravity platform behaviour and dowel forces during installation on the sea bottom. In B. Mazurkiewicz, editor, *Offshore Platforms and Pipelines - Selected Contributions*, pages 217–298, 1987.

[101] G. Mesri and Y. Choi. Settlement analysis of embankments on soft clays. *Journal of Geotechnical Engineering*, 111(4):441–464, April 1984.

[102] G. Mesri and Y. Choi. The uniqueness of the end-of-primary (EOP) void ratio-effective stress relationship. In *International Conference on Soil Mechanics and Foundation Engineering*, volume 2, pages 587–590, 19985.

[103] M. Miner. Cumulative damage in fatigue. *Transactions of the American Society of Mechanical Engineers*, 103(67):159–164, 1945.

[104] Y. Mostafa and M. El Naggar. Response of fixed offshore platforms to wave and current loading including soil-structure interaction. *Soil Dynamics and Earthquake Engineering*, 24(4):357–368, 2004. doi: 10.1016/j.soildyn.2003.11.008. URL `http://www.sciencedirect.com/science/article/B6V4Y-4BVRSWB-1/2/e5713bb57db1fb0daa8b7c21cec76c72`.

[105] S. Murayama and T. Shibata. On the rheological characteristics of clay. Technical Report 26, Disaster Prevention Research Institute at the Kyoto University, Kyoto, Japan, October 1958.

[106] W. Musial, S. Butterfield, and B. Ram. Energy from offshore wind. In *OTC - Offshore Technology Conference*, 2006.

[107] NGI - Norwegian Geotechnical Institute. Bearing capacity of gravity platforms on sand: Calculation procedure. Technical Report 52422-2, NGI - Norwegian Geotechnical Institute, 1988.

[108] NGI - Norwegian Geotechnical Institute. Bearing capacity of gravity platform foundations on sand: Static and cyclic laboratory tests on very dense Sand. Technical Report 52422-5, NGI - Norwegian Geotechnical Institute, 1989.

[109] A. Niemunis. A visco-plastic model for clay and its FE-implementation. In *11th Colloque Franco-Polonais*, 1996.

[110] A. Niemunis. *Extension to the hypoplastic model for soils*. PhD thesis, Schriftenreihe des Institutes für Grundbau und Bodenmechanik der Ruhr-Universität Bochum, 2003.

[111] A. Niemunis and I. Herle. Hypoplastic model for cohesionless soils with elastic strain range. *Mechanics of Cohesive-Frictional Materials*, 2:279–299, 1997.

[112] A. Niemunis and S. Krieg. Viscous behaviour of soil under oedometric conditions. *Canadian Geotechnical Journal*, 33:159–168, 1996.

[113] A. Niemunis, T. Wichtmann, and T. Triantafyllidis. A high-cycle accumulation model for sand. *Computers and Geotechnics*, 32:245–263, 2005.

[114] A. Niemunis, T. Wichtmann, and T. Triantafyllidis. Long term deformations in soils due to cyclic loading. In W. Wu and H.-S. Yu, editors, *Modern trends in geomechanics*,

volume 106 of *Springer Proceedings in Physics*, pages 427–462. Springer, 2005. doi: 10.1007/978-3-540-35724-7.

[115] F. Norton. *The creep of steel at high temperatures*. Mc Graw Hill Book Company, Inc., New York, 1929.

[116] K. Nübel. *Experimental and numerical investigation of shear localizations in granular material*. PhD thesis, Publication of the Institute of Soil and Rock Mechanics in Karlsruhe, 2002.

[117] Nysted Havmøllepark. Velkommen til Nysted Havmøllepark, 2009. URL `http://www.nystedhavmoellepark.dk`.

[118] Offshore Princess Amalia Wind Farm Project Management. Prinses Amalia Windpark, 2009. URL `http://www.prinsesamaliawindpark.eu/en/index.asp`.

[119] H. Poulos. Cyclic stability diagram for axially loaded piles. *Journal of Geotechnical Engineering*, 114(8):877–895, August 1988. doi: 10.1061/(ASCE)0733-9410(1988)114:8(877). URL `http://link.aip.org/link/?QGE/114/877/1`.

[120] H. Poulos. Cyclic axial loading analysis of piles in sand. *Journal of Geotechnical Engineering*, 115(6):836–852, 1989. doi: 10.1061/(ASCE)0733-9410(1989)115:6(836). URL `http://link.aip.org/link/?QGE/115/836/1`.

[121] H. Poulos. Pile behaviour - Theory and application. *Géotechnique*, 39(3):365–415, 1989.

[122] L. Prada Sarmiento. Comparison of two constitutive models to reproduce the behaviour of granular soils under cyclic loading. In ISSMGE, editor, *3rd South American Conference of Young Geotechnical Engineers*, 2009.

[123] M. Rahman, H. Seed, and J. Booker. Pore pressure development under offshore gravity structures. *Journal of Geotechnical Engineering Devision*, 103(12):1419–1436, December 1977.

[124] M. Randolph, J. Cassidy, S. Gourvenec, and C. Erbrich. Challenges of offshore geotechnical engineering, state-of-the-art paper. In *Proceedings of the 16th International Conference on Soil Mechanics and Geotechnical Engineering*, pages 123–176, 2005.

[125] D. Rebstock. Hypoplastic simulation of piles and column foundations. In H. Brandl and F. Kopf, editors, *16th EYGEC - European Young Geotechnical Engineers Conference*, pages 303–312. Österreichischen Ingenieur- und Architekten-Verein, 2004.

[126] D. Rebstock. Kleinbohrpfähle unter statischer und zyklischer Belastung - hypoplastische Simulationen. In J. Stahlmann, J. Gattermann, and M. Fritsch, editors, *Pfahl-Symposium*, pages 349–365. Institut für Grundbau und Bodenmechanik der TU Braunschweig, 2005.

[127] A. Sawicki and W. Świdziński. Compaction curve as one of basic characteristics of granular soils. In E. Flavigny and D. Cordary, editors, *4th Colloque Franco-Polonais de Mechanique des Sols Appliquee*, volume 1, pages 103–115, 1987.

[128] A. Sawicki and W. Świdziński. Mechanics of a sandy subsoil subjected to cyclic loadings. *International Journal for Numerical and Analytical Methods in Geomechanics*, 13(5): 511–529, 1989.

[129] A. Sawicki, W. Świdziński, and B. Zadroga. Settlement of shallow foundations due to cyclic vertical force. *Soils and Foundations*, 38(1):35–43, March 1998. URL http://ci.nii.ac.jp/naid/110003946117/en/.

[130] A. Scheuermann and H. C. On the feasibility of pressure profile measurement. *IEEE - Transactions on Instrumentation and Measurement*, 58(2):467–474, 2009.

[131] H. Seed. Soil liquefaction and cyclic mobility evaluation for level ground during earthquakes. *Journal of Geotechnical Engineering Devision*, 105(2):201–255, 1979.

[132] H. Seed and I. Idriss. On the importance of dissipation effects in evaluating pore pressure changes due to cyclic loading. In G. Pande and O. Zienkiewicz, editors, *Soil Mechanics - Transient and Cyclic Loads*, pages 53–70, 1982.

[133] H. Seed, K. Mori, and C. Chan. Influence of seismic history on liquefaction of sands. *Journal of the Geotechnical Engineering Division*, 103(4):257–270, 1977.

[134] C. Słomiński. *Validierung von Rechenmodellen zur Scherzonenentwicklung mit Versuchen im Labor und in situ.* PhD thesis, Publication of the Institute of Soil and Rock Mechanics in Karlsruhe, 2007.

[135] Sonderforschungsbereich 398 "Lebensdauerorientierte Entwurfskonzepte unter Schädigungs- und Deteriorationsaspekten" in Verbindung mit der Pressestelle der Ruhr-Universität Bochum. RUBIN Wissenschaftsmagazin - Sonderheft 2009 zum Sonderforschungsbereich 398, 2009. URL http://www.ruhr-uni-bochum.de/rubin/sfb398/index.html.

[136] H. Sturm. Modelling of biaxial tests on remoulded saturated clay with a visco-hypoplastic constitutive model. Technical report, Institute of Soil and Rock Mechanics in Karlsruhe, 2004.

[137] H. Sturm. Numerical simulation of centrifuge tests of an offshore gravity platform on very dense sand with an hypoplastic constitutive law. Master's thesis, Institute of Soil and Rock Mechanics in Karlsruhe, 2005.

[138] H. Sturm. Shallow Foundations of Offshore Wind Turbine Structures in the North and Baltic Sea. In *18th European Young Geotechnical Engineers Conference, Ancona (Italy), 2007*, Ancona, Italy, June 2007.

[139] H. Sturm, O. Solf, and P. Kudella. Selbstheilung von Offshore-Gründungen für Windenergie-Anlagen. In T. Triantafyllidis, editor, *Grundlagen und Anwendungen der Geomechanik - GKK 08*, Publication of the Institute of Soil and Rock Mechanics in Karlsruhe, pages 153–170. Institute of Soil and Rock Mechanics in Karlsruhe, 2008.

[140] H. Sturm, O. Solf, and P. Kudella. Self-healing effects of shallow foundations for offshore wind turbine structures. In *11th Baltic Sea Geotechnical Conference*, 2008.

[141] M. Taiebat and Y. Dafalias. SANISAND: Simple anisotropic sand plasticity model. *International Journal for Numerical and Analytical Methods in Geomechanics*, 32(8):915–948, Juni 2008. doi: 10.1002/nag.651.

[142] C. Tamagnini, G. Viggiani, and R. Chambon. A review of two different approaches to hypoplasticity. In D. Kolymbas, editor, *Constitutive modelling of granular materials*, pages 107–145. Springer, 2000.

[143] M. Topolnicki. *Observed stress-strain behaviour of remoulded saturated clay and examination of two constitutive models*. PhD thesis, Publication of the Institute of Soil and Rock Mechanics in Karlsruhe, 1987.

[144] D. J. und Grabe J. Monopilegründungen von Offshore-Windenergieanlagen - Zum Einfluss einer veränderlichen zyklischen Lastangriffsrichtung. *Bautechnik*, 85(5):317–321, 2008.

[145] T. Utumi. Untersuchung zur Verformungsentwicklung von Flachgründungen unter Offshore Windenergieanlagen typischer Belastung für das Langzeitverhalten mit Hilfe eines Akkumulationsmodells. Master's thesis, Institute of Soil and Rock Mechanics in Karlsruhe, 2009.

[146] K. Valanis and J. Peters. An endochronic plasticity theory with shear-volumetric coupling. *International Journal for Numerical and Analytical Methods in Geomechanics*, 15 (2):77–102, 1991.

[147] R. Verdugo and K. Ishihara. The steady state of sand soils. *Soils and Foundations*, 36 (2):81–91, 1996. URL `http://ci.nii.ac.jp/naid/110003946012/en/`.

[148] G. Voyiadjis. Kirk C. Valanis - A biographical sketch. *International Journal of Plasticity*, 22(8):1393–1397, 2006. doi: 10.1016/j.ijplas.2005.12.001. URL `http://www.sciencedirect.com/science/article/B6TWX-4J3WS8S-1/2/1b4241d219960f8315ee341503cca5c6`. Special issue in honour of Dr. Kirk Valanis - Valanis Issue.

[149] R. Whitehouse. *Scour at marine structures*. Thomas Telford Publications, 1998.

[150] T. Wichtmann. *Explicit accumulation model for non-cohesive soils under cyclic loading*. PhD thesis, Schriftenreihe des Institutes für Grundbau und Bodenmechanik der Ruhr-Universität Bochum, 2005.

[151] J. Wiemann, K. Lesny, and W. Richwien. Gründungen von Offshore-Windenergieanlagen - Gründungskonzepte und gotechnische Grundlagen. Technical Report 29, Mitteilungen aus dem Fachgebiet Grundbau und Bodenmechanik der Universität Essen, 2002.

[152] P.-A. v. Wolffersdorff. A hypoplastic relation for granular material with predefined limit state surface. *Mechanics of Cohesive-Frictional Materials*, 1(3):251–275, 1996.

[153] W. Wu. *Hypoplastizität als mathematisches Modell zum mechanischen Verhalten granularer Stoffe*. PhD thesis, Publication of the Institute of Soil and Rock Mechanics in Karlsruhe, 1992.

[154] T. Youd. Compaction of sands by repeated shear straining. *Journal of the Soil Mechanics and Foundations Division*, 98(7):709–725, July 1972.

[155] M. Zaaijer. Comparison of monopile, tripod, suction bucket and gravity base design for a 6 MW turbine. In *Offshore Windenergy in Mediterranean and Other European Seas (OWEMES conference)*, April 2003.